GW01339442

Before Columbus

Other books by Samuel D. Marble

Guide to Public Affairs Organizations

Glimpses of Africa

Before Columbus

The New History of Celtic,
Phoenician, Viking, Black African,
and Asian Contacts and Impacts
in the Americas before 1492

Samuel D. Marble

South Brunswick and New York: A.S. Barnes and Company
London: Thomas Yoseloff Ltd

©1980 by A.S. Barnes and Co., Inc.

A.S. Barnes and Co., Inc.
Cranbury, New Jersey 08512

Thomas Yoseloff Ltd
Magdalen House
136-148 Tooley Street
London SE1 2TT, England

Library of Congress Cataloging in Publication Data

Marble, Samuel Davey, 1915-
 Before Columbus.

 Bibliography: p.
 Includes index.
 1. America--Discovery and exploration--
Pre-Columbian. I. Title.
E103.M36   970.01   78-75321
ISBN 0-498-02370-2

PRINTED IN THE UNITED STATES OF AMERICA

To
Gladys
who said:
May I help?

Contents

	Acknowledgments	9
	Introduction	11
1	The Testimony of Columbus	23
2	The Celtic Essence	28
3	Celtic Migrations	45
4	Tools for Discovery	61
5	Utopian Mystery	72
6	Land of Persons or Prisoners	103
7	The Skill of the Phoenicians	109
8	The Wisdom of the Egyptians	117
9	The Mayan Synthesis	129
10	From Mastery to Oblivion	154
11	Montezuma's Testimony	168
12	Vikings: Moody Adventure	173
13	History and Mystery	183
14	Problems of Early America	211
	Notes	224
	Bibliography	238
	Index	241

Acknowledgments

I am indebted to the many authors and scholars whose works are reported, analyzed, and credited within this volume, and to those at A. S. Barnes and Company whose painstaking and meticulous editing brought this book to completion. To Louise Jackson and Mary Anderson I wish to express my appreciation for the preparation of fine drawings and illustrations. Dover Publications has given permission to quote extensively from the works of Hjalmar Holand. The cheerful and indispensible assistance of my wife in working on the manuscript made this undertaking a pleasure.

Introduction

This book marks a point of reassessment on a journey which began thirty years ago and which is still underway and whose destiny is unknown. With a group of students, I sat on the floor until the early hours, and in the course of discussion someone mentioned an improbable person who had vast, if not absolute power, and who had combined power with a belief in brotherhood and peace. The man was Ikhnaton, Pharaoh of Egypt, who had a queen of rare beauty. That evening marked the beginning of the study of the 18th Dynasty whose closing years surely overlapped those of Moses who grew up in the court of the Egyptian King.

After reading this capsule of Egyptian history, I realized that an understanding of the truly revolutionary thought of Ikhnaton and his Queen could be appreciateed only within the whole sweep of Egyptian history from the time of Menes, the first Pharaoh, to the time the empire was extinguished by Rome. As I began to broaden my base of understanding, I had the thrilling experience of becoming "literate," and was reminded again how little we know of another people unless we understand their language. In the presence of all the symbols of the Egyptian culture, I was an outsider until I took the time to study the hieroglyphs and let the Egyptian voices speak directly.

A knowledge of the language of the ancient Egyptian began to awaken an awareness of how much the culture of these people had permeated the present day and how far the echo of those voices had been heard. It was a surprise, for example, to discover how many words presently in the English language were in the standard Egyptian vocabulary of 3,500 B.C. and are virtually unchanged in meaning. I had read the wrong books, for no one had told me that the English language owed an indebtedness to the Egyptians.

History tells us little about the travel of individuals or of mercantile missions that circled the globe. Whether carried by ship, by camel, by horse, or in a peddler's pack, Egyptian goods were transported to the southernmost points of Africa. Bronze castings made in Egypt prior to 1750 B.C. are found on the banks of the Zambezi in Southern Africa, and on the Island of Madagascar.[1] Egyptian beads and bracelets have been unearthed as far north as Cologne. Rings and jewelry of Egyptian craftsmanship managed to find their way across the entire continent of Asia to China and, perhaps, became a token of admiration from some Chinese traveler to his beloved.

This concourse was not a one-way street. By one hundred years after Christ, China was the principal supplier of ceramics and pottery to Africa. So voluminous was this traffic that Chinese pottery fragments remain to this day one of the principal methods of dating archeological finds, not only on the coast of East Africa but for a considerable distance inland as well. Chinese products were known in Egypt and apparently were exchanged for gold and ivory. The Egyptians gave, and they received as well.

We know when the Iron Age came to Egypt. The first known implement of iron was an import—the blade of a dagger found on the mummy of the Pharaoh Tutankhamen, as shiny and rustless today as it was when it was first placed next to the body of the boy King. This was the only piece of iron in that fabled tomb and it must have been regarded as a treasure to have been placed in the innermost wrapping next to the mummy itself. The Egyptians had recognized the superiority of iron weapons, but credit for the discovery of iron-making is given to the Hittites who ushered in the new age approximately forty years before the death of the young ruler.

Three hundred years after the first iron objects were introduced to Egypt similar iron implements were welcomed on the other side of Africa and had crossed the entire distance of the continent below the Sahara to the Atlantic Ocean. How this transmission of material took place we do not know. Iron implements may have been carried across the Sahara by hand, or they may have been transported by boat, or passed on by boat and land from the Mediterranean. Nevertheless, this perfectly remarkable act of communication did take place, although no one knows by whom or when or how.

After the beginning of the Christian era, an even more remarkable event, the Christianization of Africa below the Sahara, occurred. This missionary enterprise later faltered, and masses of people were reconverted by the Moslems, leaving cultural pockets where traces of the Egyptian language survive today from the coast of Zaire in the west to the Coptic church of Ethiopia in the East. We have documents that trace the evangelical spread of Christianity in Africa, its curious challenge by the slave trade and Mohammedanism, and its subsequent dissolution and virtual disappearance.

At a somewhat earlier time, the Nestorian Christians were subject to persecution by the Orthodox Church and departed from their ancient homeland, as have countless other Christian pilgrims, to find a place where they could worship God in peace and freedom. From Egypt and the eastern Mediterranean, they trudged an incredible journey of 10,000 to 12,000 miles to China where they were welcome because of their high moral character and administrative skill. They became a trusted elite in the administrative corps of the Emperor of Peking.

Man, an animal with inordinate curiosity about other men, is not only curious but also flamboyant, rash, foolhardy, brave, bigoted, uncommonly shrewd, resourceful, and, when he wishes to be, virtually indestructible. Because these qualities are distributed in some degree through all mankind, men have penetrated into every corner of the world, and some have settled where they were; some have pushed on to other shores; and some have returned to report what they have found.

I was aware of the gifts of ancient Israel to our present age. The legacy of Greece was known by educated men as the world's first democracy where the fires of discovery burned with incandescent heat. Our indebtedness to Rome—organizer, civilizer, administrator—was understood. But far less known was our debt to Egypt, whose 3,500 years of literate history was still a smudge in the collective memory of mankind. And a discovery of this indebtedness leads to awareness of our silent and unspoken debt to many other races, other peoples, other climes.

And thus my study of life along the Nile became the study of the animating ideas that took root in that soil and then moved on and were disseminated to the four corners of the earth. The study of one people became the study of all people bound together in an irresistible urge to know each other, and who in spite of tragic and horrendous breakdowns in the process, are sharing what they have and what they know.

We have all become a part of each other—and unless we elect the option of mutual obliteration, the forces leading to a brotherhood of man will continue to enrich us and disturb us. The introduction of new ideas may result in the virtual breakdown of some societies as they have in the past, and new civil wars may be ahead of us. However, we must increasingly acknowledge our indebtedness to all mankind for what we are now, for those concepts we value, and for those tools we use to build the present and to shape the future. We are indeed children of a common past.

If we are not surprised that Egyptian castings found their way to South Africa before 1750 B.C., or that Chinese ceramics were traded on the coast of Africa at the time of Christ, we should not be surprised if other individuals with the same inquisitive and acquisitive characteristics might find a way to the Americas. The ceramic manufacturer, it should be remembered, was reasonably sure of his market before he started his product on its way. His cups and saucers had already traveled 3,000 miles when they reached India. His trade, which expanded massively in the first century after Christ, had another 1,700 miles to go to reach the eastern coast of Africa if his ships headed across the open seas on the monsoon winds, or 2,500 miles to go if the ships hugged the shores down to the Zambezi River. If such traffic were possible on the east coast of Africa, why is it not plausible that the same drive would break over the 1,520 mile ocean barrier that separated Cape Verde from Brazil?

America had been discovered at least once 8,000 years before Christ by people who presumably crossed the frozen Aleutians and then walked into every habitable crevice of the two continents from Alaska to Tierra del Fuego. Among the traces of these ancient people are found the remnants of other races who discovered these same shores and who brought, among their intellectual baggage, a knowledge of other languages, other folkways, and other worlds. As we study these evidences of a cultural mix, are we looking at false clues or is this authentic evidence? We view the colossal sculpture of an African head with unmistakable Bantu features carved from basalt and left in the jungles of Mexico and say this is evidence of an African presence. But can we be sure when we have no names, no records, and no reasons?

Our history as Americans is far more complex, far more exciting, and far more valuable than most of our teaching has led us to believe. It begins before the arrival of Columbus and the waves of migration that followed him. And although we may never know the name of a traveler from the West who put foot on these shores before Columbus saw San Salvador on

October 12, 1492, we know that ideas comparable to European, Mediterranean and North African grew and luxuriated in these lands, and we neglect them only to our own impoverishment.

One other discovery along the way was an awareness that there are those who resist, even resent, a suggestion of our interdependence, and particularly on those people whom it is now stylish to dislike or to disregard. They deny that any indebtedness has occurred, and can prove that no borrowing of skills or ideas has ever existed. They appear to have a vested interest in maintaining an ethnic and cultural purity unencumbered by "alien" influence. Some of them may be scholars who have worked out their theories on the source of the human race, and have a stake in keeping these theories intact. Others, possibly for economic or political reasons, appear to be troubled by evidences of human kinship and the sense of obligation that kinship implies. As a result, there is always a dispute as to the validity of the evidence of human transactions; and since one new fact always raises at least two new questions, it is unlikely that any conclusion will ever be final.

The fact that ceramic chips are found along 2,000 miles of the African coast is evidence that the Chinese carried on some form of trade with the east coast almost from the days of Christ. Or is it? Is this evidence sufficient to prove such trade actually occurred? The fact that Egyptian glyphs are scattered through the carved monuments on Mayan buildings still being uncovered in the tropical thickets of Central America is evidence that Egyptian learning at one time reached these shores. Or is it?

Botanists have studied the totora reed which grows only in fresh water and at one time luxuriated along the banks of the River Nile. This plant also grows in the fresh waters of Lake Titicaca of Peru. Because the totora reed propagates itself below the level of the water, botanists can think of no way this reed—which is genetically identical to its African counterpart—could have found its way to the Andes unless it was carried by hand. Is the totora reed evidence of an African presence?

On the occasion when totora reeds were woven into bundles to make a boat for the River Nile, men wore birdlike hats with beaks. Such men are shown in ancient Egyptian carvings and in paintings on jars. And men wearing hats with beaks are shown on ancient tracings in Peru where boats are being woven from totora reeds in a pattern almost identical to that of ancient Egypt. Reeds, boats and hats—are these a cultural anomaly? Is this social combination a billion-to-one anthropoligical accident? Is there something about reeds that gives rise to boat building, and something about boats that gives rise to beaks? If the answer is "yes," then it is not necessary to assume that all these unique and complicated social inventions were conveyed as a package by a visitor from overseas. However, any explanations of how reeds give rise to boats, and boats to beaks, would be just as esoteric as an explanation of how ceramic chips can be found along stretches of Africa if we reject the notion that Chinese craftsmen shipped these objects over 6,000 miles of land and sea—a millennium and a half before Columbus.

So we look at evidence that is fastidiously gathered together and try to determine what it means. The rulers of the ancient Inca Empire were of different genetic stock from the Indians they governed. On the basis of what we know of herdity and human evolution, there is no conceivable way these tall, white, bearded redheads could emerge from the genetic reservoir of American Indians. But if they were not part of the first migration to American shores, they may have been part of a later migration. But from where, and when, and why?

Certainly this one volume does not contain more than a sampling of the data that have been gathered in the last ten years, or the last fifty. Some of theories of human enlightenment that involve direct revelation or extraterrestrial intervention are not mentioned here. Neither are they rejected, but rather are left to the evaluation of individuals who are better qualified for such a task than I. This book is confined to the observable and reportable, and to the shrewd surmises that can be made on the basis of what is already known and verified. This information is agonizingly and infuriatingly inadequate. But new findings come to light each season, and in weighing this information the reader is invited to use his own scales.

Books on American history start with 1492. When Columbus landed, American history began. From that time on we know from whence Americans came and why. They came from the Spanish vigor tinged with greed for gold; from the English search for religious freedom; from William Penn's gentle settlements in Philadelphia and North Carolina that sought a peaceful accommodation with the Indian; from the French who brought us metropolitan thought tinctured with the slightest trace of piracy; from the hard-working Germans who felled the forests from Ohio to Wisconsin, and from rebels for freedom like Karl Schurtz; from the Scandinavians who broke the plains of the Midwest. These were our ancestors, and it was from stock like this that we came. At a still later date, Italy and Germany sent us those victims of tyranny with great minds like Enrico Fermi and Albert Einstein. From this stuff American character was forged.

The fact that we are now expanding our consciousness to include our inheritance from the blacks, the Indians, and the Mexicans has the effect of involving us with a wider planet, but also with the future of all humanity. Furthermore, if Christopher Columbus was not the first European to visit America, we know this fact is going to be hard to digest as was Copernicus' discovery that the earth was not the center of the universe—a discovery that created havoc during the fifteenth and sixteenth centuries. In spite of the evidence in view around us, we have continued to cling to the primacy of Columbus as one item of our historical faith which comforts us by giving us some fixed boundary marks. Now those boundaries, too, are crumbling and much of this new thinking is the product of the last ten years. Evidence of previous exploration has been in existence for some time. Information derived from archaeological sources and from scientific research has clearly expanded the parameters of American existence. New documentation has been found, and massive efforts at the translation of codices and fragments of documents have brought new light. Most important is the shift in the willingness of the American to look again at his own history with the same sense of honesty and awareness that Copernicus had when he faced the fact the the earth followed a course around the sun.

The change can be illustrated by citing one of our great historians, Samuel Eliot Morison. Not only a full professor at Harvard, he was also an admiral in the Navy. He received a glittering list of awards and decorations, and in 1965 produced *The Oxford History of the American People.* As a product of the Oxford University Press, this was a definitive work. In it, Morison devoted exactly fourteen pages to America before the arrival of Columbus. Ten years later, however, he completed two volumes—both as fat and handsome as the earlier Oxford work—on *The European Discovery of America, 500-1600;* from fourteen pages to 1,400 pages!

During the years to come, American history will become far more complicatged and fascinating than most of us had ever suspected. In the process of discovering who we are, we will make some startling discoveries about the achievements of our American forebears in which we can take enormous pride—as well as some events in which we will have an unavoidable sense of sadness, and of shame.

Before Columbus

1 • The Testimony of Columbus

AMONG those who wondered whether Columbus was the first person to discover the new world was the man Columbus. He recorded in his journal a visit with King Juan of Portugal, who told him before his departure that he, the King, had evidence of trade between Africa and South America in tobacco, cotton, shell money, and bread root for a considerable time in the unknowable past.[1]

Columbus entered in his log that when he arrived at Haiti the natives informed him that six European expeditions had arrived before him.[2] He was unable to evaluate the significance of this information because Haiti is only one of the Antilles Islands that stretch for 1,500 miles from Florida to Brazil. It is entirely possible that six different expeditions, each like Columbus', set out from the East, and each one "accidentally" landed at Haiti. It would appear more probable that Haiti, once discovered, had systematically been charted, mapped, and had become a recognized port of call.

Columbus brought back reports of commercial transactions between South America and merchantmen from across the sea. Negroes sold spear heads to the Indians made from a metal called "guanin." The native Indians were unable to identify the minerals of which their weapons were made, or explain how they were forged or hardened. They did agree that the source of the product was trade with Negroes, and that this trade had existed over an extended past. Columbus assumed that "guani" came from Guinea in West Africa. He brought samples of the material home for assay by Ferdinand's own mineralogists, and their conclusion was explicit: "The very alloy is of African origin."[3]

On his second voyage, Columbus began a process of island-hopping that brought him closer to the mainland and flushed out his knowledge of the chain of Antilles he discovered on his first voyage. As he traveled this course, two things impressed him: the unquestionable Negro presence, and the apparent influence of Christianity. He put these two social observations together and came to the conclusion that he might be on the trail of Prester John, Emperor of the Christian Kingdom of Ethiopia. When he reached Cuba, his convictions on this matter were sufficiently confirmed that he launched an expedition inland to search for this fabled king![4]

What, you say? Prester John on the Island of Cuba? How ridiculous! How could Columbus, a man who was right about so many things, become so hopelessly befuddled?

In Columbus's behalf, we need to remember that in 1493 he was not aware that he had discovered a New World. He still believed he was on the coast of Asia. When he reached Cuba, he was under the impression that he had found the continent, and he required all the members of his crew to make an affirmation that they had discovered the mainland, so that on this point the expedition had unanimity.

At this time, the location of Ethiopia was wrapped in massive obscurity. Some reports made Prester John a prince more spendid in wealth and dazzling strength than any prince in Europe. His kingdom was somewhere in Africa or, as some believed, in Asia. As Columbus put together the clues he found on each island, he may have begun to dream of a discovery that permitted a diplomatic liaison between Christianity in the West and Christianity in the East, with the infidel in between.

On his third voyage to America, Columbus began his travel to the New World by sailing south to Guinea, Africa. Before departing, he visited again with King Juan about the African-American trade and about the reported voyages from Portugal to the lands. Juan gave him some highly useful instructions, predicting that if he were to follow a more southerly route he would reach the mainland. Columbus followed these suggestions and redeemed these prophecies by finding the coast of Venezuela.[5]

How much more did Don Juan know that he did not tell? The Portuguese, like the Phoenicians, maintained an iron secrecy about their line of trade and sources of supply. Don Juan had already experienced a bitter battle with the Spanish Crown over how far Portuguese hegemony extended into the Western Hemisphere, and this struggle finally had been settled by the Pope. He displayed little interest in feeding the Spanish King's hunger for foreign provinces. With the door firmly but politely shut in his face, Columbus decided to sail first to Africa to confirm directly what Don Juan had predicted, and what he also suspected about the volume of African-American trade. Because of his African stopover, he was not only able to plot a voyage that actually brought him to the shore of South America but also to learn in Guinea something more important than anything else he had turned up on his third voyage to America: that the French, Portuguese and Germans had gotten to Guinea before him.

As a subject for research, the possibility of African discovery of America has never been a tempting one for American historians. In a sense, we choose our own history, or more accurately, we select those vistas of history for our examinations which promise us the greatest satisfaction, and we have had little appetite to explore the possibility that our founding father was a black man for whom two continents should be named, as well as great nations, rivers and cities. In spite of the remarkable collection of clues assembled by no less a person than Columbus that other sailors had preceded him to America, and that an extensive commerce of African manufacture had been carried on for years through well-defined channels, this chapter of American history has proportionately gone unnoticed and unexplored. And not this chapter alone. In general, the view that the American continents had been explored before 1492 is not one that evokes Yankee enthusiasm. Neither is the possibility that America had been settled by Vikings for five hundred years before the arrival of the little Spanish fleet. Here the motivation was curiosity on the part of the Scandinavians. It was they who found the time to dust off the records and reconcile them with the sagas. It was the descendants of the Vikings who provided labor to dig out stones of an original settlement on the banks of North America. Over the span of years, the Scandinavian evidence became so overwhelming that the defensive historian, Admiral Samuel Eliot Morison, found it necessary to concede that the Greenlanders did precede Columbus to these shores by half a millennium.[6] It was a reluctant conversion but it was an honest one. However, the search of the North European anthropologists for evidence of settlements has only started. More is to come.

In comparison to the Scandinavian studies, the knowledge of African expeditions to the Americas has not moved a great deal beyond where it was when Columbus left it. However, some stimulating work has been done and provocative inquiries have been made. According to Leo Weiner in *Africa and the Discovery of America*, a first tentative, touching of America by native Africans in their own boats may have come sometime after A.D. 1100.[7] The presence of the French and presumably the Portuguese on the Guinea coast closest to America in the century before Columbus probably signaled a rapid expansion of trade in South America—the same trade that Columbus discovered, verified, and reported to his king. Batallia-Reis lists between fifteen and twenty expeditions from Africa westward between 1436 and 1500. And antiquarians have unearthed three charter grants of land made by the Portuguese King, John II, to his subjects between 1484 and 1486—grants that would permit them to establish colonies, and would seem to be quite meaningless if there was no knowledge that land existed in the Western Hemisphere.[8]

So we have real records to deal with and more will come to light. Additional evidence will appear with the unearthing of artifacts, foundations, building stones, and articles of trade on this side of the Atlantic. The search has been started and will continue. Curiously, the most provocative findings have not come from the Americans but from the Swedes, Norwegians, Australians, Irish, and others. We can now review this evidence and ask ourselves the question that was surely in the mind of Columbus. If there were earlier visitors to these shores, who could they have been?

2 • The Celtic Essence

THE Celts were among the first. They may have arrived here as early as 800 B.C. with the rise of their culture in Europe. A second wave may have occurred in A.D. 300 to 500 as an expression of their religious exploration. The third wave may have brought them here as refugees from religious persecution. And then after Columbus, there were more and more.

The Celts were the "born losers" of history. At one time the Celts—or Gauls—dominated Europe. For six centuries they battled the Romans. They fought bravely, fanatically, with no fear of death. For six centuries they retreated with stupendous casualties, but with a colossal and sovereign contempt for Rome. They passionately hated all that Rome came to represent: organization, mechanization, materialism, hunger to control others, and cruelty.

In the year A.D. 410 when Rome was rich, fat and corrupt, it was toppled by the Celtic Christian, Alaric; but victory was fleeting, and soon the Celts were on the defensive and were pushed back into the crevices and corners of Europe—Ireland, Wales, Scotland and Brittany. There they continue to resist the modern successors of Rome.

Who were the Celts?

Over the centuries, those who study them say: They believe more that reality is within than without. They are a people who refuse to surrender. They refuse to be treated as masses. They are individuals to the end. They know they are part of a universe, and religion is to find their place within it. They are irrepressible, joyous, winsome, and melancholy. They confuse poetry with fact, and are often confused as to which is more important. They are the ultimate defenders of the person, which to them is the one eternal and everlasting essence.

They continue to be the "losers" of history. But while we know what they have lost, we do not know what they have won. Perhaps within each one of us there is something that is still part of our Celtic inheritance; and in the growing struggle be an individual, the Celt within us may be our savior.

Barry Fell of Harvard, who is both an oceanographer and epigrapher, has discovered a substantial number of Celtic inscriptions in the New England area. Among the more important are the antiquities of "Mystery Hill," a site which, Fell is confident, was occupied from 800 to 600 B.C.[1] Mixed in with the Celtic writings are stone-engraved tablets of Phoenician origin, possibly from Cadiz, which became one of the major shipping centers of pre-Christian Spain.

Also found in the New England area are caves which are constructed from stone obviously gathered from a considerable distance to form a structure not otherwise found in colonial New England. These structures have corbeled walls that slope to a narrow roof which allows one stone at a time to be removed from the ceiling to permit a view of the sky, a construction used in Celtic Europe primarily for purposes of astronomical observation. Stone structures of corresponding construction are found in Ireland and on the Iberian peninsula, but from the earliest time they have been known in the New England states as "root cellars." The owners of the property on which such caves are located have assumed that someone in previous generations, a grandfather or a great-grandfather, constructed these caves for storage purposes, although there appear to be no records to show that any of the ancestral owners of such a property ever constructed the "root cellar," or that it was not there when the property was originally acquired. Experts on colonial architecture insist that the corbeled roof design has no precedent in colonial architecture, and the suggestion that these unique chambers were constructed by colonial settlers is untenable.

Archaeologists have been slow to accept Fell's analysis, pointing out that to date no skeletons have been found that could be used effectively to date the structures and that no other Celtic artifacts have been unearthed, although significant pre-Columbian artifacts have been found at distances both near and remote from the shelters that Fell has identified in such fascinating manner in his book, *America B.C.*

While pre-Columbian artifacts have been found to exist, it is worth noting that an American Stonehenge was constructed near "Mystery Hill," and the stones placed with such precision as to mark accurately the equinoxes, and the winter and summer solstices. This arrangement could not have happened by accident, and carbon-14 analyses of charcoal found near these sites at different layers indicate that occupation prior to 1000 B.C. is evident. Reports on carbon-14 dating are now on deposit in the Widener Library at Harvard University.[2]

While Celtic artifacts have not been found in the corbeled structures themselves, and while little excavation has been done in the outlying sites, nevertheless bronze weapons have been found in mound builder sites dating back to 200 B.C. or earlier, as well as other tools and implements, which are accurately fashioned after European implements that were used at approximately the same time in Spain and on the Iberian peninsula. Bronze, of course, is a mixture of copper and tin, and tin is not found in North America. Therefore, these implements were either imported from Europe—probably from Spain—or American craftsmen learned how to make these implements based on European models with the knowledge that they would have a dependable supply of tin available to them on a commercial basis.

Dozens of Celtic inscriptions in Ogam writing are also found in the United States; but there is the possibility that these could be forgeries, could be mistranslated, and some are weathered by time so as to be virtually unreadable and therefore unreliable. The authenticity of these inscriptions can only be established by searching for and finding other cultural artifacts and evidence of their origin that make their credibility and dating possible. One such related cultural feature is that many of these inscriptions found on stones are referred to as phallic symbols.[3] As a matter of fact, many of these stones are not symbols. The portrait of man is not a symbol of the man but a portrait. Normally we think of towers, smokestacks, campaniles, and Maypoles as phallic symbols. These rocks are carvings of the phallus itself, and should be regarded as phallic sculptures.

Inscribed stones identical in character are found in Ireland and Spain, and in various places in the United States. In Mexico, there are phallic stones, each with its Ogam writing, similar in appearance to those found in County Kerry, Ireland. At Uxmal, Mexico, these stones, each with Ogam writing, each from four to five feet high, each placed in rows, stand very much like fence posts. It is here that the Empress Carlota, wife of Emperor Maximilian, perhaps meditating on her childless estate and "knowing that her own days of lovemaking were over," wandered and wondered while her husband took an Indian mistress to sire a son.[4]

Ogam script dated about 800 B.C. has been found in the West Indies. Inscriptions in Phoenician script, presumably left by sailors from the Port of Cadiz, have been found in Brazil and have been dated at about 538 B.C.[5]

Carved phalluses are somewhat awesome to consider. Presumably of pre-Christian origin, the dating of any stone monument which has been exposed to two or more millennia of seasonal weather is an uncertain undertaking unless the carvings are accompanied by other cultural clues. One of the most impressive figures of this sort is known as the "Cerne Giant," which is cut into the chalk bedrock of a hillside in Britain.[6] The giant sculpture portrays a naked warrior 180 feet high; customarily, the Celts entered battle naked. The Cerne Giant has a phallus 35 feet long and has been described as the "largest phallic monument in Britain."

The phallic sculptures both in Ireland and England, as well

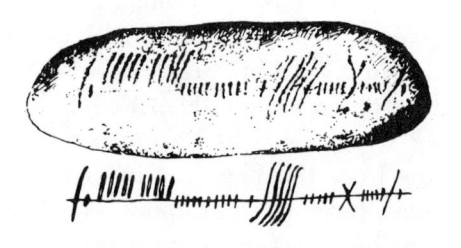

Approximately 400 phallic stones with ogam marking have been found in England. On this side of the Atlantic, comparable stones have been found in New England, Mexico, and Venezuela. The stones discovered in Europe are of pre-Christian Celtic origin, and the stones found in the Americas presumably are the same. Most of these stones have been exposed to severe weathering, and any person who has visited a graveyard with interments in the eighteenth century is aware of how 200 years can blur an inscription—to say nothing of 2,000 years in the case of the Celtic markers. However, the discovery in the same vicinity of corbeled chambers and the premeditated arrangement of stones for ritualistic purposes argue for the authenticity of these findings.

as in the United States, Mexico, and Brazil, are accurate in detail in spite of 2,000 years or more of erosion. As would be expected, the phallus is portrayed erect with utter fidelity, including the prepuce and glans. Carved of stone, these sculptures are hard, angry, passionate, and demanding. The fact that they bear Ogamic writing makes them appear to be strangely scarred, as if the possessor had notched each of his conquests on the organ, just as the frontiersman notched his rifle for each Indian warrior slain.

The fascination with the phallus is referred to daintily as a "fertility rite," but of course a wedding is also a fertility rite and therefore the terminology fails to reveal the Celts' passionate preoccupation with this member. Fascination and admiration were attached in a way that appears to have religious overtones, and it would not be an exaggeration to say that at one time among the Celts there appeared to have been a "cult" of the phallus.

Ogam script was used by the Celts until about A.D. 800 when it fell into disuse, possibly because of religious persecution. However, no more than a most cursory study of it was attempted for years until Barry Fell stirred a tempest by collecting a set of matching inscriptions from parts of the world bordering on the Atlantic Ocean. As an oceanographer, his study of ocean currents made him aware that movements of water provide a relatively rapid and reasonably predictably highway for moving from continent to continent. These currents often have dramatic effects on climate, as when the Gulf Stream strikes the European sea coast near Bergen, Norway. He reasoned that these currents might carry with them not only changes in meterological conditions, and a variety of ocean fauna and flora, but might be the ocean conduit for the transportation of man and his cultures. As he set out to test the relationship of the ecology of man to the movement of the ocean, one of his liveliest discoveries was the presence of Ogam script at relatively predictable points as indicated by the flow of ocean currents.[8]

Ogam script did not become a concern of historians until late in the present decade. Anne Ross, who is undoubtedly one of the world's leading scholars in Celtic history, dismisses Ogam writing with a passing reference in her book *Pagan Celtic Britain*, published in 1967. The author is indebted to Anne Ross for much of his knowledge of the character of Celtic culture, but she dismisses Ogam, saying merely that it is "a strange alphabet" and worthy of separate study.

What we know about Ogam script is, very briefly, as follows: Edward Lhyd of County Kerry, Ireland, wrote a letter in 1712 to the Royal Society in London reporting that he had found a stone near Dingle in County Kerry[9] with curious writing on it. He had no idea of the source or the meaning, but he did send a reproduction of the inscription to be kept in the Society's files for future reference.

That same year, Cotten Mather, the well-known Puritan clergyman who had a powerful influence in casting American Christianity in its ascetic mold, also wrote in 1712 to the Royal Society to inform them that he had discovered strange writings on a rock at Dighton. The information of Mather's discovery was also carried in the transactions of the Royal Society.

For the next seventy years, the interest in Ogam writing lay dormant. During that period, the American colonies separated from Britain and established their own scientific societies and their own centers for the study of history and the antiquities. It was in the year 1784 that an English colonel, Charles Vallancey, reported that he had discovered an ancient tombstone in County Clare, Ireland,[10] and explained how he had deciphered the text on this stone. He called the inscription "Ogam," and recognized that it was identical with the unknown script Lhyd had discovered three generations earlier. Vallancey had scored a stunning achievement, which was noted in England but in this country was apparently ignored.

The most significant finding reported by Vallencey was that he had found his own "Rosetta Stone," known as the *Book of Ballymote*, which gave a sampler of Ogamic writing on several pages and parallel to it the corresponding Roman lettering. This discovery opened up a new world of understanding; and although the significance of this achievement was understood slowly in England, nevertheless over a span of years following Vallancey, more than 400 Ogam monuments which had been erected prior to the Roman invasion were found and identified in Britain south of Hadrian's Wall.

In the passage of time, the knowledge that Cotton Mather's tablet was now susceptible to translation managed to cross the ocean, and Alec Reinert, a linguist on the faculty of Harvard and a specialist in South American languages, found a rock inscription he could not read on a cave wall in Paraguay.[11] After careful search, he discovered that the record was in Ogamic writing, and reported a visit by mariners from Cadiz about 300 to 500 B.C. The discovery of two Ogamic inscriptions on this side of the Atlantic brought a lively revival of interest in this ancient form of writing.

The relationship of the Ogamic writing of the Irish and the use of a comparable script by the Iberians in Spain will be explored in somewhat greater detail in the latter part of this chapter. Let it be noted here that the Ogamic writing of the Celts in Spain may have influenced the Punic script of the Phoenicians who built a trade center at Cadiz. Whether such an epigraphic transfusion took place or not, nevertheless a stone with Punic writing was found at Moundsville, Grave Creek, West Virginia in 1838, and it is believed that it has now been correctly translated. Another stone bearing the Aton sun symbol of the Pharaoh Ikhnaton, engraved in approximately 800 B.C. in Punic wriitng, has been found in Oklahoma with words that appeared to be excerpts from

Ikhnaton's hymn to the sun.[12] Further examples of Ogamic writing in this country continue to be found, and Dr. Barry Fell has established a center and a journal for the reporting of such discoveries.

The Celts occupied that portion of the European continent which the Romans referred to as Gaul. The Romans were at war with the Celts constantly for eight centuries, from the earliest rise of Rome until the assault of Rome by the Celtic chief, Alaric, in A.D. 410. Roman culture reached its peak in terms of territory controlled in A.D. 350, after which it was necessary to withdraw troops from Britain to defend the retreating Roman forces on the European continent. As the legions of Rome withdrew, the Celts took their place.

The Celts were a fearless and high-spirited people. They rushed into war as if they were rushing into games. A Celtic baby was first fed with food from the tip of a sword—a kind of baptism into the faith that the child should eventually die in battle.[13]

The Celts were not only warlike—they were vain. They loved jewelry. Men and women were fond of fashion, clothes, ornamentation, and display. They loved drinking, particularly on celebrations and holidays. The largest wine krater—a wine mixing bowl—ever found was thirteen feet in diameter, now exhibited in the Museum of Chatillon sur Seine.[14] Their alcoholic revelries must have been stupendous. Considering the size of the largest wine bowls used by Celts, it would be possible for a guest literally to be drowned in alcohol—perhaps a more blissful way to pass into the netherworld than through the pain of battle.

The early Celtic religion was pantheistic. Their worship was chiefly out-of-doors; they were part of and partners with nature. All of the living world around them was a divine revelation, and they worshipped at streams, wells, springs, and running water. The spring had a special fascination for them, because they considered it a passage to the underworld and the afterlife.[15] They dropped coins, jewelry and other precious objects in springs, in wells, and in running water, and from this comes our tradition of throwing coins in the fountain.

Over 3,000 holy wells are still in existence in Ireland today.[16] Most of them are dedicated to saints, but pagan rites still linger. Sunrise circuits are made of wells in various sections of the country, and at each station it is customary to drop something into the well—a coin, or something as insignificant as a piece of cloth or a pin. Each well is considered to be endowed with express magical properties and can transmit aid to the pilgrim, in the from of a cure for rheumatism, toothache, or stomach pains. Certain wells may improve fertility or give added strength to those who need it. The wells may answer the wishes of him who wants to love and to be loved, to grow in understanding, and to live to a ripe old age.

The "Cult of the Head" was central to Celtic life.[17] Heads were ubiquitous, even in the *Book of Kells*, which is a Christian production and is on view at Trinity College in Dublin. This tradition continued into the eleventh century, and indeed it continues to the present day. It fuses through our entire culture to the point that it is no longer recognized as Celtic in origin. The head on covers helps to sell magazines; the head validates coins and gives them numismatic value; it sells billions of dollars' of cosmetics; women constantly change the size and shape of glasses, and whether vision is improved may be irrelevant if the head is improved; enormous wealth is sluiced into beauty parlors. Let these modern Celts give thanks.

The Celtic warrior rushing naked into battle would detach the heads of his victims and carry them home as trophies.[18] If he rode as a charioter, he would hang the heads around the neck of his horses. Sometimes he would nail the skulls to the outside of his house for reasons of vanity and also as an exhibition of his prowess. As indicated, a skull was considered an appropriate drinking vessel. The portrayal of many heads on monuments, sometimes coming from a common neck and occasionally representing as many as five or six heads in a cluster—like grapes—was an aritistic and religious tradition which we now refer to as "polyencephaloi."

Contrary to what one might expect from preoccupation with the phallus, the Celts were anything but male chauvinists. Or perhaps it would be accurate to say that the phallus was regarded with great esteem by both men and women. "Women's liberation" was an actuality under Celtic law, which accorded women a position of equality. In marriage they continued to own all that they brought to the marriage; and if in such a combination the woman were the possessor of greater goods and wealth than the man, she became the head of the household. Women often accompanied their husbands into battle, and achievements of heroic proportion are linked with the names of women warriors. A famous woman, Boudicca, led the attack and burned Roman London in A.D. 60. In one of the early American colonies, a woman was credited with saving the settlement single-handed when in approximately A.D. 990, a group of Norwegians settled on the coast of Newfoundland. For a period of two or three years the Norwegians lived amicably with the Indians. Gradually, however, they began to cheat the Indians until the relation between the settlers and the indigenous people became hostile. Unexpectedly, the "Skraelings" (wretches), as they were known to the Norwegians, launched a sudden and pitiless attack in which the settlers were outmanned and which appeared to portend the end of the settlement itself. At the height of the attack, when the future seemed to be most precarious, Freydis, who was recorded as a "very beautiful and most exceptional woman in every respect," seized a sword, tore off her clothing to bare her breasts, and ran out into the open, beating a sword against her breasts and screaming like a hellion.[19] This dramatic appearance so startled the Skraelings that they were totally immobilized. Silently they withdrew and never appeared again.

With their lusty love of life, the Celts in Ireland had a poetic vein that gave a quality of charm and winsomeness to their literature which has an appeal for us that is undiminished today. They were impulsive, but they were intuitive and mystic. Their literature ran the gamut from unrestrained delight to stark disaster. It was they who gave us the legend of King Arthur and his Court—a court of men united at a round table in a search for the good: the Holy Grail. But it was a court in which there was a lapse of fidelity. Celts gave us the story of Tristram with his doom-begotten love for Isolt of the purple eyes, who moved as through a dream to the inevitable tragedy of their mutual destruction.[20] The cadences of this love still sound in the music of Richard Wagner and the poetry of Edwin Arlington Robinson.

What the Celts gave us down through the centuries is hard to separate from all the rest of our inheritance, because all of us to some degree are now Celts. But we do know they gave us the Christmas tree. We know that long before the birth of Christ, the Druid climbed to the top of the oak to cut with a golden knife the mistletoe that hung in the hall and endowed those who stood beneath it the capacity to be loved and to love in return. We know they gave us the festival of May Day, in their calendar the first day of the year, which they called the "Day of Beltane," and the Maypole to be wound with colored ribbons accompanied by dancing. They gave us Halloween, which in spite of its name, was never effectively assimilated by Christianity and its theologians. Halloween brings back for at least one day the Celtic world of witches and goblins, ghosts and spooks, black cats and jack-o'-lanterns. They gave us "coins in the fountain," the legend of fairies, and the desire to make a wish on a shooting star.

But this is not all. Much of what the Celts gave us touches

deeply on religion, and it is in this realm that we find an explanation for their early voyages to the American shores.[21]

They were "madly fond of war," the Greek historian Strabo says of them. Their love of warfare made them natural mercenaries, and they were fearless and venturesome ones, too. Because of their nomadic tradition, they traveled great distances to fight for pay. They were hired by Egyptian pharaohs as early as the third century B.C. At the request of the King of Bithynia, they were invited to cross the Bosporus at Byzantium into the area of present-day Turkey. For fifty years they continued almost uninterrupted warfare in behalf of the Bithynians, but they were finally stopped by the King of Pergamon and settled for a time in an area that was recognized as an independent province, and which bore the name "Galatia." They maintained their political independence of Rome; but as Roman pressures on them grew, they passed back again into Europe and continued their travels, which led them at least in token force into what is modern-day Russia. In addition, and more important, they traveled north and westward in Europe outside the Roman Empire.

During the days of Galatia, the Celts were visited by the Apostle Paul who effectively established a number of lively churches. A natural affinity appeared between Christianity and Celticism, and Paul's letter in which he instructed these early churches is contained in the New Testament as *Galatians*.[22] He refers to them as "foolish Galatians," and cautions them against "the observance of days," which would be particularly appropriate for the Celts who did not follow a religious calendar of seven days' duration, but rather observed the solstices and the change of the seasons. They had observances for midwinter and midsummer, which gave Shakespeare a chance to write a play full of Celtic fantasies as part of *A Midsummer Night's Dream*.

All during this period of time, the Gauls, or Celts, were at war with Rome. While Christianity could be spread inside the Roman Empire, as it was, there was virtually no way for Roman Christianity to be transmitted outside the Empire to the enemies of Rome, and this became increasingly true as Christians entered the Roman army, and as Christianity became an official religion of Rome. Not until the fall of Rome in A.D. 410 at the hands of the Celts did this situation change.

There was one exception, however, and that was the Galatians. They were Celtic Christians—enthusiastic, life-loving, irrepressible Christians—and they were outside the Empire. They were the enemies of Rome; they were united with the rest of Gaul in tradition, spirit, and particularly in having a powerful and dangerous common enemy bent on their total subjugation. And so the Celtic Christian Church, outside the Roman state, traveled along the edges of the Empire into Germany, the Scandinavian countries, and into Ireland. The pace of this missionary effort must have been one of considerable celerity because the Celtic Christian Church was well established in Ireland by the end of the second century A.D., considerably before Augustine appeared to introduce Christianity in England.

The Celtic Christians were Arian.[23] Arius, who gave his name to this belief, was a theologian and a leader, born and educated in Alexandria, Egypt. He intensely influenced the Eastern Church, which came to accept the Arian point of view as its own. It was natural that the Celts—living among Arian Christians—should absorb this theology. Arianism was a doctrine of Unitarianism or Monotheism, in contrast to Trinitarianism, and was largely confined to the seminaries until Constantine the Great recognized Christianity as an official religion of the Roman Empire. At this point, the division between the Arians and the Trinitarians began to assume painful proportions.

Constantine wrote to the leaders on both sides of the Arian

division objecting to the dispute and urged that the matter be dropped. He declared that he was basically unable to understand the subtle theological distinctions. He considered the controversy primarily one of words, with no ideas of substance involved. During his lifetime, when he moved the headquarters of church and state from Rome to Constantinople, he appeared to be both on one side and then on the other in this struggle of ideas; but it was in a Celtic Christian Church that he was first "converted" and saw the writing in the sky, "In hoc signo vinces." When he was finally baptized on his deathbed, he was baptized by an Arian bishop.

Arius himself lived to be excommunicated by one of the ecumenical councils of the Christian Church. Ten years later, a second ecumenical council reversed the decision of the previous one and restored Arius to good standing. Meanwhile, Constantine had passed away and his son, Constantius, had become Emperor and was head not only of the Holy Roman Empire but of the Holy Roman Catholic Church. Constantius was an Arian Christian, and when he came to the throne Saint Jerome was to observe, "The world woke up to find itself Arian."

With the passage of the centuries, it may be difficult for the modern reader to understand why the world could be thrown into turmoil over a theological question, or to understand very clearly what the theological differences were. In this respect, at least, he would be a fellow-traveler with Constantine the Great. Arianism insisted on monotheism and held that if it were possible for God in the form of a "Spirit" to impregnate the body of a woman and make her son into a God, it would be possible also for the deity to enter other women and create other gods, and thus open the way back to polytheism.

After the death of Constantius, the Arian-Trinitarian controversy continued until it was formally terminated by Theodosius,[24] the first Spaniard to become His Imperial Majesty as the head of the Holy Roman Empire. Theodosius was a man of unusual intolerance and cruelty. Once during a brawl in the hippodrome in Thessalonica, a member of the Imperial Guard was killed. In retaliation, Theodosius ordered that 7,000 people be gathered at random from the city of Thessalonica and taken into the hippodrome. The gates were locked, and with the Imperial Guard in command the 7,000 were slaughtered. It was with this kind of determination that Theodosius set out to "neutralize," or "purge" the Arian bishops and Arian theologians of influence. When he felt he had the ground sufficiently cleared, he called the Ecumenical Council at Constantinople in 381. The Nycene Creed with its affirmation of the Trinity was re-established. Arians were declared heretics, and were forbidden to assemble. Their churches were destroyed and they lost their right to possess and inherit property. However, Theodosius II went further than his predecessor. He ordered immediate death for anyone who denied the Trinity, and all Arians were to be executed without delay and to the glory of God.

From the insistence on the Trinity as the only Christian position sprang other heresies that, in common with Arianism, had problems with the Trinity. One of these was the Monophysite position, and four of the Orthodox Catholic churches separated from Rome following the Council of Chalcedon. Another heresy was that of the Nestorians. Specifically, the Nestorians believed that Mary was the mother of Jesus but that Mary was not the mother of God. Their logic found fault with the Trinitarian idea that God was able to quicken the womb of Mary and thus become a son of his own seed. They believed that it was sacrilegious to think of god as nuzzling at the breast of Mary, or to think of Mary as wiping his bottom. To question the relationship of Mary to God was to invite persecution, and under the threat of death the Nestorians were extruded from the Holy Roman Empire as rapidly as possible, all the way to Mongol Empire of China where they were welcomed and where several mothers of

Khans were Christians. As late as the year 1277, the leader of the Mongol Embassy to England was an Nestorian Christian.

All the time the Arian-Trinitarian struggle went on inside the Holy Roman Empire, the Celtic Christians were outside that Empire, and either they knew very little about the theological turmoil or, if they did, felt unmoved. Rome had always been their enemy; Rome threatened to destroy them as Celts; and now Rome threatened to destroy them as Christians. So nothing was changed. They carried their religious faith to the Celtic kings of Germany, the kingdoms of France, northern Spain, and of course to Ireland.[25]

Arianism was the fundamental difference between Celtic and Roman Christianity. However, there were other and seemingly trivial differences as well.[26] Roman monks, for example, shaved their heads and wore a tonsure around the edge of their hat somewhat like a halo. It was claimed that this reflected the crown of thorns that Christ wore at crucifixion. On the other hand, Celtic monks shaved off all their hair except a topknot, the latter being no doubt an adaptation from the Druids, who followed a similar practice.

The two churches followed different calendars, which brought their observation of holy days into conflict. Easter came at a different time. While the Celts were fasting, Romans were feasting; and while the Romans were fasting, the Celts were feasting. Because the "forms" of Christianity had a powerful influence on the thinking of early Christians, these differences, which seem small, created disturbances that were enormous.

3 • Celtic Migrations

THE fusion of Celtic culture with Arian Christianity was a heady mixture. The buoyancy of the Celts joined with the freedom from structured dogma provided a release—almost an explosion—of energy. The Arian Christians, who had been on the defensive and had been repressed and rejected, suddenly found themselves free of Rome with the entire non-Roman world to missionarize and conquer. The Celts, possibly because they had been enemies of Rome for eight solid centuries, greeted Arianism with excited expection. Freed from system and centralized control, and with an option to speak to local needs and traditions, Christianity was propelled across Europe at record speed, and may have reached Ireland by the end of the first century, and was well developed by the second.[1] By the fifth century, Celtic Christianity made Ireland one of the cultural centers of Europe. In his book *Civilization* Lord Kenneth Clark depicts Ireland, presumably while the rest of Europe was immersed in the "Dark Ages," as the one repository of learning and culture. For four centuries, namely, from approximately A.D. 400 to 800, they preserved the learning of the ancients. From some source, possibly from St. Patrick, but more likely from St. Paul himself, they retained the classical Latin language which across Europe had broken down into a variety of vernacular that eventually produced the potpourri of Romance languages, all aberrations of Latin.[2]

Celtic Christianity traveled across northern Europe, then gradually began to seep south with the crumbling Roman Empire; and in the transformation, Ireland became the spiritual home of the Celtic Christians. They lacked the centralizing influence of a Holy City, a pope, and a well-defined hierarchy. Christianity in Ireland was decentralized into several hundred monasteries,[3] and the primary missionary effort went into the establishment of new monasteries, or communities, which extended as far east as the cities of Regensburg, Germany, and Kiev, now part of modern Russia, and whose extension into westward will be referred to in subsequent pages.

As Rome slipped into moral decay, so did the control of the Roman provinces. Colonialism had never been popular with the colonized; even in Israel where the Romans permitted a high degree of independence by allowing the Jews to have a king of their own with a degree of self-government, the Romans were hated. Even more so to the north where the Romans exercised direct control, people were chafing under Roman tyranny. Taxes had gone up; the quality of Roman justice had gone down. Dishonesty, bribery and extortion had become a recognized part of each dealing the Celt had with his Roman conqueror. Furthermore, a substantial part of the army of occupation had become mercenaries; and while their nominal loyalties were to Rome, their personal interests were geared to their own advantage. Some of the mercenaries were drawn from other occupied areas under Roman control and thus they may have had an unmentioned and unconfessed sympathy with the very people under their domination.

The distaste of the Gauls for Roman control and their bitter resentment of Roman corruption found expression in a very extraordinary man who came from one of the strongest of the Celtic tribes, the Balthi, who dwelled along the lower Danube. Alaric, who had served in the Roman army and had withdrawn with disillusionment and disgust, became immediately the universal choice of the Celts as their leader. He was a Christian, a Monotheist, and a spokesman of resistance to Roman control.[4]

Celtic armies were never managed with the machinelike precision of the Roman Legions, and therefore success in battle depended far more on leadership than it did on command.

Alaric's first strike was at the heart of the Empire in Constantinople. When he realized he would be unable to take the city, he retreated to Greece, investing three years in the liberation of those people from Roman control before he retreated beyond the Alps.

Alaric made the first of three moves against Rome itself in A.D. 400. His army, accompanied by wives and children—Alaric's wife among them—moved down the Italian peninsula to begin a siege of Rome. As the Easter season arrived, they suspended fighting and placed their defense in the hands of Eternal God expecting that their fellow Christians in the Holy City of Rome would be respectful of their desire to reverently observe the crucifixion and resurrection. However, Alaric's trust in the sanctity of Easter as immunity from attack was in vain, and Stilicho, general of the Holy Roman Empire, fell on the Celts with murderous attack. A large portion of the Celtic forces was destroyed, and it was reported that Alaric's wife was taken captive as well. However, Alaric did not withdraw his forces at once, and as a price for returning north of the Alps persuaded the Roman senate to make concessions that would lighten the burden of Roman occupation. Alaric had no sooner withdrawn from Italy in 403 when the Romans reneged on their promise and Roman occupation became more agonizing than before.

The second siege of Rome began in 407, when the Gauls under Alaric's command surrounded the city, cut off food and water, and quickly brought the city to negotiation. Alaric as a Christian was determined that a change in the Imperial system could be brought about without bloodshed, and his demands on the Roman people were essentially modest. He did not ask for a dismantling of the Empire, but rather for an easing of central control.

The senate of Rome made concessions demanded by Alaric which included not only a restoration of Roman justice but also a protection of otherwise defenseless people from the brutalizing ravages of Roman soldiers who as occupying forces were able to live outside the law. Alaric asked religious toleration for Celtic Christians. He proposed that he himself be appointed a regent of Rome for his own Danubian province which lay strategically between Rome and Constantinople. Rome agreed to these demands and Alaric again withdrew his army beyond the Alps without bloodshed. But no sooner had the Celts withdrawn than Rome reneged a second time and introduced a new wave of Celtic Christian prosecution. The life of the occupied people became little better than that of animals.

The third approach of Alaric and the Celts occurred in 410. Because Rome was dependent for bread and water from sources outside the walls, the siege was a brief one. Before Alaric entered the city, he gave instruction that any person who sought sanctuary in a Christian church—whether he was Roman Christian, Celtic, or pagan—should be spared. No buildings were to be destroyed unless employed for military purposes. Large supplies of bread were accumulated to feed the people immediately after the siege was lifted. During the fall of the city, there was considerable looting—not by Celts but by Romans—and once order was restored, the Celts sought out the gold and silver vessels and other ecclesiastical paraphernalia and returned them to the respective churches.[5]

Even such destruction as occurred during the third siege could have been avoided had not Theodosius II in Constantinople urged the Romans to make no concessions and to continue resistance. He promised reinforcements if the Romans would hold their ground, and sent thirty legions of men, so great was his hatred of the Celtic Christians. But all these efforts proved ultimately in vain. After the fall of the city, Alaric continued his march to the foot of Italy, but on the way he died of fever. The Celts marched on, crossed the Mediterranean, took control of the African province—the bread basket of Rome—and held it for more than a century.

Historians invariably speak of Alaric as a "barbarian," but he was a Christian who very nearly lost an entire army because he stopped his military operations long enough to observe a religious fast. He knocked on the gate of Rome three times, not asking for warfare but for an easing of the most painful burdens of colonialism. His cause was curiously liberal, 1,500 years ahead of its time in urging restraint and religious toleration in the face of the threat of death by Theodosius I and Theodosius II—two of the most cruel and powerful Christians of any time who were already laying the basis for the Inquisition and the future religious wars of Europe.

Historians for 1,600 years have been referring to the "sack of Rome" in 410. Historians will probably continue to do so for 1,600 years more. The term "sack" is not only an exaggeration but also a seriously unfair description of what was done by a man who was far more compassionate than the Romans. Rome was not "sacked"—certainly not in the Roman sense. When Rome defeated Carthage, the Romans "sacked" that city—not one stone was left standing on another. More than a million people—soldiers, noncombatants, women, children and aged—were murdered. Refugees in flight were sought through the plains and hills and put to death. The fields—hundreds of thousands of acres—were sown with salt. When Rome "sacked" Carthage, Carthage died forever. But when the Celts moved out of Rome, most of Rome was left standing, and some of the same Rome stands today.

When the forces of Alaric moved into Africa, they brought with them Celtic Christianity which from the year 410 began a rapid spread along the entire area of the southern coast of Mediterranean Africa. Catholic churches converted to Celtic Christianity, as well as did the Donatists. Celtic Christian monasteries were established, and church scholars agree that over this span of more than a hundred years, the Roman

Church came close to disappearing.[6] However, in 533, Justinian recaptured the African colonies for the Holy Roman Empire and the Christian Celts again became refugees. As the Holy Roman soldiers advanced along the coast of northern Africa, large numbers of Celts entered Morocco and hid in caves. From there they moved on to whatever security they could find. Some of them became pilgrims to new lands across the seas—to the land of the Americas. The record of their persecution, their flight, their plan to migrate to the lands beyond the oceans, as well as the record of those who had gone before them was written on the walls of caves in the Atlas Mountains where it remains today.

During the remainder of the fifth century, most of the kings in the vast area that was known as Gaul became Arian Christians, including Spanish, Iberians, Burgundians, Visigoths, and others, but not the Frank known as Clovis. Clovis remained pagan until 496. If Clovis had accepted Celtic Christianity, a confederation of Arian Christians would have occupied the center of Europe, including portions of Spain, Ireland, and probably the Scandanavian peninsula. However, as Roland H. Bainton says in *The History of Christianity*, "Clovis did not want a confederation. He wanted a conquest." He passed directly from paganism to Roman Christianity.

"I cannot endure that these Arians should possess any part of Gaul," he said, " and with God's aid, we will go against them and conquer their lands."[7] However, while no one would question the sincerity of Clovis's conviction, his conversion gave him strategic advantages for conquest, bringing the support of all other Roman Christians to his side. By unifying the ranks, Clovis provided a highway for the Roman Church's militia to move to the north and on to Britain, Germany, Scandanavia, and Ireland. The day before Clovis's conversion marked the high point of Celtic Christianity in Europe.

Clovis's move into the Roman fold again placed Celtic

Christianity in Europe on the defensive, and its decline was swift and inevitable. A few battles, a twist of fate, a bit of connivance, a touch of luck, and Europe became Roman Christian rather than Celtic Christian. Rome put its strong stamp on the Christian message. Looking back, it is difficult to view the religious tapestry and see the unique Roman threads which were woven into Christianity so firmly that today it is not possile to separate them. Today, we are all children of Rome even though we may have reached our faith through the heretical routes of Lutheranism, Calvinism, or Henry VIIIism. From this point in time, it is not possible to tell how history might have changed if these ingredient events had been different. However, it is possible to look at ourselves today and understand that some of the basic distinctions between Roman Christianity and Celtic Christianity still exist, and to know why some ambiguities survive within us, and to glimpse how our society might have been if the balances of history had differed.

The contribution of Rome to Christianity was great. Rome transmitted its science of administration to the Christian Church along with the Roman arch, great roads, and great armies. Essential to Roman administration was a strong hierarchical system disciplined at every level to function with a speed that was virtually reflex action. At a later stage in the Empire, this well-honed discipline would begin to dull as the number of mercenaries increased and as the opportunities for corruption multiplied. Lewis Mumford describes how the precision of military organization, with its emphasis on the commitment of total energy toward the achievement of a single goal, how the uniform established the concept of "the replaceable part," and how the effectiveness of mass action became the model not only for Christianity but for industrialism in the eighteenth century.

The Roman legions were the finest piece of human machinery yet devised. Because each man in the organization could be trusted to function with almost electronic efficiency, a small handful of men could achieve victories that were otherwise denied to enormous masses. Julius Caesar, for example, never commanded more than 35,000 men. At the end of his Gaulic wars, he reported that he had fought thirty battles, obliterated 800 towns, and had killed 1,982,000 men, women and children. This is efficiency! With administration, organization and hierarchy, one Roman soldier was able to kill approximately sixty people. And the two million men, women and children who were killed were Celts, whose defeat was due not to lack of courage and armament, but to lack of organization. The Celts fought as individuals; the Romans fought as machines of 3,000 to 6,000 men.

Along with organization and administration, Christianity absorbed the military values on which the Roman state was built. The taking of life in war was unanimously condemned by all Christian leaders in the period prior to Constantine the Great. They held that the act of taking life was incompatible with the injunction of the Scriptures "to love one's enemies." According to Roland H. Bainton of Yale University, all second- and third-century Christian writers from Tertullian of Carthage to Origen of Alexandria believed that the triumph of Christianity would so change the quality of society that war would vanish.

Of all the characteristics of Celtic culture the most profound was the deep sense of individualism. For the Celt, the idea of any army based on the pattern of a Roman Legion would have been unthinkable. The complete submergence of the person to forge together a legion that became the tool of one man would have been slavery, whatever other benefits might have been derived. It was this rejection of Roman organization and Roman hierarchy which permeated Celtic thought, and which became the special distinguishing characteristic of Celtic Christianity. But to describe the strong features of Celtic Christianity is also to describe its weaknesses.

Unlike Roman counterparts, the Celtic Christians were not church-centered. It was not their practice to gather in closed structures, but rather at springs, the source of rivers, waterfalls and wells. Their pilgrimages were to mountaintops—places rich with reverie and reminiscence. Their church calendar was one that focused on changing seasons—the solstices and the equinoxes, the birth of the year, the dying of summer, the pulse of life, the coming and going of a time, the subtle web that contained man's mystic relationship to the universe.

The Celtic Christians were not priest-oriented or preacher-centered. Their faith in immortality was as strong as that of any people since the time of the Egyptians. However, they did not need the offices of the priest to travel from this world into the next. "Druids" were their wise men, their judges, their collective memory, their men of virtue, their arbiters of disputes. They had few priestly functions. Their role was far more that of teacher than performer of rite and ritual. Because the druid was a conveyor of knowledge, the Irish Saint Columba could say that "Christ is my druid," meaning that "Christ is my teacher," and implying that it was not impossible for a druid to be a Christian.[8]

The Roman practice was to extend Christianity by building and creating churches and by requiring attendance. The Celtic pattern was to establish monasteries and create communities. The Irish monasteries were institutions which gave intellectual and cultural leadership to Europe for a period of almost two hundred years. The Celtic monasteries became not only centers of worship but also educational institutions and producers of books, Bibles and treatises, including the classics of Plato and Aristotle. They were manufacturing centers as well. For more than a century, the finest ceramics in Europe were manufactured in Celtic monasteries. The quality of Celtic metallurgy was unsurpassed. Garments, fabrics, cheeses, and utensils produced in the Irish monasteries exceeded in quality those manufactured anywhere in Europe, and thus the Celts gave expression to the belief that all of life should be incorporated in Christianity, and that the difference between the sacred and the secular, the holy and the profane, the lay and the priestly, should be greatly eased—if not almost erased. In all the structure of Celtic Christianity stretching from Ireland to the mouth of the Danube, there was no pope, there were no cardinals, there were no bishops or archbishops. Each community had its abbot and therefore enjoyed a high level of self-sufficiency. It is difficult to conceive how under such a decentralized form of church government anything like a unified creed could come into existence. The Celts were individualists in their belief and in their practice. As a notably intuitive people, they had more confidence in the spirit of a living Christ to unite them than they could ever have had in organization, a creed of words, or superb administration.

Both the Celtic and the Roman Christians seemed to love a fight—contrary as this may be to the teachings of the founder of their faith. Part of the Roman inheritance was an insinuation into Christianity of cruelty, and this on an organized basis meant inquisitions, extinctions and eradications.

To the Celts, conflict was a personal affair; to the Romans, it was corporate. Had Celtic Christianity predominated, we might have had more hilarious fistfights, and even a larger number of murders, but fewer wars. In Roman Christianity there were fewer violent personal confrontations, but there was more warfare; for according to Quincy Wright in *A Study of War*, Christians during the past three hundred years have been more warlike than people of any other faith or profession.

Ireland presumably was converted by St. Patrick.[9] Fortunately, we have a brief biography from his own pen which describes how as a Britain he was trained in a monastery on the continent and with the title of "bishop" went to Ireland in 432. There his own account ends, and what happened over the next century is a matter of myth and mystery. The one thing that must have struck him with particular force when he arrived in Ireland was that the Christian Church had already preceded him without blessing from Rome, and that a large number of well-functioning monasteries were in existence. Whether Patrick's faith underwent any alteration we do not know, but we are sure that he launched his missionary effort from the city of Armagh from a cathedral which remained non-Roman, a building rich with Celtic symbols.

Neither does it appear that Ireland was Christianized in Patrick's generation, because a century later St. Columba or Columcille, like Patrick, was later beatified by Rome, but during his lifetime he was unquestionably the spokesman for the Celtic Christian Church.[10] And as he moved down from the north, his efforts confronted those of Augustine moving up from the south, and eventually Christians were converting Christians. Late in the seventh century, King Owsy instructed the Arians to accept the Trinitarianism of Rome, and by the end of the eighth century all knowledge of the use of Ogam writing had disappeared.

Migration by the Irish Celts to the west and to the shores of North America occurred in two waves, the first reaching a peak about 300 B.C. as part of the expanding missionary movement during the period of dynamic religious and cultural development on the Emerald Isle. The second and greater wave may have begun early in the 600s and increased consistently into the tenth century A.D. Although the code of Theodosius had been amended by Justinian so that the death penalty was no longer mandatory for non-Trinitarianism, that was all that had changed. The Arians were still heretics. More specifically, they were pagans. They had no rights under Roman law or under church law. It was not only proper to separate them from their possessions and their inheritance, it was also legally required of Roman Christians to do so. The forces were gathering to give expression to that unique form of

Christian activism known as "The Inquisition." Celts in Spain were feeling the hot sting of Christian hatred even more than the Irish. While Celtic church structures constructed in Spain still exist today, it was becoming abundantly clear, even before the end of the first millennium, that there was safety only in Orthodoxy or in flight. As we look at the traces of Celtic writing, we should remember that pilgrims to the shores of North and South America may not only have been Irish and Spanish; they may also have been pilgrims seeking escape from English pressures in the highlands of Scotland, in Wales, or in Cornwall and Brittany.

No one knows just when the Faroe Islands were discovered, but when the Vikings arrived the Irish monks were already there. No one knows just when Iceland was first discovered, but coins cast in London in 257 have been found in its excavations. When the Vikings arrived to claim the Greenland territory, the Irish monks were there, and Greenland is only 200 miles from the American continent.

Next to Patrick, St. Brendan, the navigator, is probably the most popular Irish saint.[11] He was born in approximately A.D. 484, and he became a famous churchman, a founder of four monasteries, and visited England, Scotland, Wales, and the continent. He doubtless dreamed of what lay beyond the western seas, and today over one hundred manuscripts exist which describe the voyages of St. Brendan. These copies are more in the form of sagas or poetry, and as in the case of the familiar sagas, the *Iliad* and the *Odyssey*, it is not easy to separate poetry from fact.[12]

What is known is this: The aged Abbot Barinth came to Brendan to report that his son Mernoc, by sailing westward, had reached the shores of "The Promised Land of the Saints." When Mernoc reached these shores, which he described as a "delicious island," he found that it was already populated with monks. Mernoc and his fellow seafarers returned again to Ireland, this time taking Fr. Barinth with them, and later returned with him to the Emerald Isle. To sum it all up, Mernoc entreated Brendan to visit "The Promised Land of the Saints."

According to the account, Brendan selected fourteen of his devoted and vigorous monks to ask them how they felt about such a calling. They agreed that, "God willing," they would follow the abbot to his death, if need be.

As nearly as they could ascertain God was willing, and after appropriate fasting and prayer they began the construction of a boat for a forty-day voyage. The boat, called a curragh, was designed like a wooden basket, covered with well-tanned oxhide and caulked with tallow. The historian Samuel Eliot Morison says that the boat "being little more than a wicker basket covered with skins, rode the waves like a cork and stayed dry in heavy seas, and which broke over heavier wooden ships." He points out that "smaller and less seaworthy vessels than this have crossed the ocean in our day."

Brendan's voyage was to the north. With his company of fourteen devoted monks, he sailed for three months, much of it through fog so dense that they could see neither sea nor sky. During this time, their provisions for a forty-day voyage ran low, and they were forced to eat only once each second or third day. Finally when the did see land, it proved to be a very high and rocky island. They had to circumnavigate the entire mass before they found the only port into which their curragh could be safely brought to shore.

Heavy fog is a frequent occurrence over the North Atlantic, and just how far they may have progressed is a matter of some uncertainty. However, one can imagine the surprise of the abbot and his fourteen argonauts when they were greeted on this craggy island by a venerable man who proved to be Ailbe, the noted Irish saint, who had founded on this island his own monastery only eight years before.

Joined by his own eleven monks, Ailbe greeted Brendan and his cohorts with hymns of welcome and by washing the feet of the newly arrived visitors "according to the scripture." The fact that Ailbe was subsequently beatified should make it possible for researchers to locate this stony island with some degree of accuracy, but to the best of available knowledge such research is yet to be undertaken.

With fresh provisions Brendan set out again, and continuing on a northerly passage moved into a "coagulated sea," apparently packed with ice. Locked in this floe, they were trapped for twenty days.

The account lacks none of the wonder of picturesque Irish courage and excitement. One day a python arose to swallow Brendan, boat and navigators, all alive. But lo! Just as the first monster of the sea was about to swallow the holy men and their boat, a second leviathan of the deep appeared and swallowed the first. At another time, a great bird dropped on the ship a branch with fruit to indicate that they were close to land. And they were. It was an island covered with "fruit, vines, and grapes which gave forth a spicy odor." However, it was not the promised land they sought, and they provisioned their ship and set forth again.

After a time, they reached an island on which they landed and kindled a fire. But the island moved, and proved to be the back of a whale that protested by dislodging the monks, who scrambled back to their curragh. Morison doubts this story. Is there any wonder why?

Eventually, after six years, Brendan and his crew reached "The Promised Land of the Saints." They visited there, worshipping with the Irish monks who had preceded them.[13]

The trip home was much shorter and was accomplished in less than a year. When Brendan arrived back in Ireland, he had reputedly reached the age of 93. Scholars believe that the date of this trip, assuming it to have been factual, was somewhere between A.D. 577 and 583. Shortly after his death, he was canonized, and then the process of enlarging the

Tim Severin followed carefully the specifications provided by St. Brendan in building his boat of leather and oak. He used identical materials for tanning and curing the cowhide. The rigging of the boat was designed to correspond to that of craft built in the fifth and sixth centuries. On the basis of his experience, Severin came to the conclusion that medieval materials were better than contemporary substitutes. To add authenticity, he ate medieval food on the voyage, and as a consequence took on a medieval smell. While the accoutrement of the voyage appeared the same, the motivation was different. Severin never had the propelling sense of mission Brendan did nor was he strengthened and sustained by the confidence that his undertaking was blessed by an eternal God. Brendan was on a voyage to extend his Father's kingdom; Severin planned his trip to write a book.

man by the addition of myth and poetry began so that the true St. Brendan is blurred by creative Irish imagination.

Quite aside from the quaint occurrences along the way, there are facts, possibilities, and plausibilities enough that it is not possible to dismiss the saga as legend any more than scholars now dismiss the *Iliad* and the *Odyssey* as folklore. After 3,000 years during which man regarded these magnificent poems as nothing more than elaborate fairy stories full of elegant imagination and charming disemblances, Heinreich Schliemann read the sagas as history and found Troy and Mycenae.

The foundation stones of the ancient Irish monasteries on the Atlantic coast have not yet been uncovered. Perhaps they will never be found, or perhaps like the ancient settlements of the Vikings they will be found only by accidental excavation under some thriving modern city. The city of Troy was unearthed after six other cities, built one after another on top of it, were carefully removed. It should be further remembered that the foundations of the colonies of Eric the Red, which were started on this continent a thousand years ago and for which more accurate records exist than those we have for the voyage of St. Brendan, were not located until this decade. In a sense, the search for pre-Columbian America has just started.

4 • Tools for Discovery

WHAT evidence can we rely on to know these things actually happened?" you ask.

The answer is there is no sure way; no evidence will convince everyone; there is no way of knowing beyond a shadow of a doubt. Those of us who have witnessed court procedure and know something of the rules of evidence know how difficult it is to establish simple facts based on eyewitness accounts. And even when eyewitness accounts are available, it is dismaying to see how many witnesses beheld the same event and saw it differently, and how many remember it differently even though the phenomenon they observed was one and the same.

While outside the realm of mathematics there are no descriptions of events that work perfectly, this does not mean that large amounts of truth are not accessible; and while it is necessary sometimes to proceed on the basis of faith or intuition, nevertheless we do know a great deal more about the universe than our fathers knew, or their fathers knew before them. Where knowledge is possible, guessing is a crime; and in the realm of human history we do know very much indeed.

Where anthropology ends and where history begins is hard to say. For most of the human adventures described in this book there is also a historical and documentary base. The growing use of scientific technique has often added evidence to confirm or contradict what our manuscripts have to say. Some of the scientific approaches to human history will be described in these pages; but in spite of all the limitations of written records, the documents, the inscriptions, the scribblings of early explorers are by far the most important key to the past and the most effective. What a tragedy it was that the library at Alexandria was burned, not so much by those who hated learning as by those who hated Alexandria. What a disaster it was that Cortes gathered up every written document, including every treaty, codex, textbook, journal and scrap of writing that he could find in the land that is now Mexico, heaped them up in front of the central pyramid in Tenochtitlan and burned them. During the years that followed, with his troops and with his clergy, he rooted out every fragment of a written document and put it to fire. With such scrupulosity and thoroughness did he do his job that only four parcels of written documents of authentic Aztec origin come down to us today. None of these four fragments can be understood more than partially. What a heartbreaking human catastrophe it is to permit one man to destroy the records of an entire civilization—complex and technically advanced, and potentially able to make a contribution to our own welfare—and to have that civilization sealed off forever from any verbal transmission to the present day.

While the records of one epoch of man's past has been blotted out, new records of man's history on the globe have been found and more are being found every day. In about 1890 in the exploration of the Royal Palace of Knossos at Crete, mounds of clay tablets were found in the basement. These cuneiform tablets had been preserved since approximately 1330 B.C. Their survival can be considered an accident. The palace and the city were destroyed by fire while the wind was blowing from the west. The valuable documents of the Royal Palace were written on parchments or on papyrus, and, indeed, all of these perished in the holocaust. However, the day-to-day activities of the Empire were recorded by stylus on inexpensive mud tablets which were dried and then filed away in the basement of the palace. These records contained accounts of troop movements, the posting of guards, the departure of ships, the sale of goods, accounts in the treasury, the payment of taxes, the records of trials and the appointment of officers and officials.

Immediately above these clay tablets in the subterranean archives in the Palace of Knossos was another great storage room in which large quantities of olive oil were kept in huge clay jars capable of holding 50 or 100 gallons. When the great fire swept the palace, these clay containers cracked and the olive oil seeped through the floor to the basement below.

Burning with intense heat, the oil fire converted the clay tablets into the equivalent of terracotta, and in this form they survived the ravages of time. Today we know very little about the court life and the treaties of the Cretan people, but we do know a great deal about the economic, commercial, social and religious conventions of the Palace and the people who inhabited the island.

The Pharoah Ikhnaton was a religious heretic who believed in one god. From the standpoint of Egypt's 2,000 other gods and their priesthood, he was doubtless insane. He built a new city as close to the geographical center of Egypt as it was possible to calculate and called it Tel El-Amarna. When he died, the priesthood placed a curse on the city and led the people back again to the capital of Thebes and to the worship of 2,000 Egyptian gods. With the passage of time, sifting sand covered over the city, and because it was cursed it was not visited again and became lost to history. Excavations on the rediscovered city were begun late in the last century, and in 1914 Knudtzon and other German scholars issued a two-volume translation of the diplomatic archives of Egypt during this portion of the Eighteenth Dynasty. These records had remained untouched until their discovery in the twentieth century, and as a consequence Tel El-Amarna is now viewed as a sort of a Pompeii of Egyptian culture, and we know more about the life of this sixteen-year period than any other comparable period of 3,000 years of Egyptian history.

It is interesting to report on the recent discovery of a

Babylonian library containing 50,000 tablets which will engage the talent of scholars for years to come in preparing translations, but the reader might also be interested to know that ancient Egyptian documents are continually turning up.

There were at least three levels of burial in Egypt up to the time of Julius Caesar and the Conquest by Rome, and the lowest and cheapest level was simply to take the body of the dead person out to the edge of the desert, wrap it carefully in waste paper—in this case, papyrus from the waste baskets of the scribes, the priests, the temples, the bureaucrats, and the army signal corp—and with appropriate prayers and supplications cover the body with dry sand. How many Egyptians were buried in this humble fashion, no one knows. The number could run into the millions. These bodies wrapped in papyrus and buried in the sands of the desert are continually being discovered, and when such a discovery is made the papyrus is carefully unwrapped and shipped to the Cabinet d' Egyptologic in Paris where translations of the writing on this waste paper are constantly being made. Much of the information being discovered is trivial in detail. However, a whole record of court cases has been found. Recipes, court orders, prescriptions, letters of condolence and congratulation, bits of scandal, and behind-the-scenes manipulations have been found, translated, and recorded. One prize found among the wrappings on desiccated bodies has been the first half of a heretofore unknown play by Aeschylus written about 430 B.C. and translated from Greek into Egyptian. Of the hundreds of plays written by Greek playwrights, only thirty-six have survived. We now have half of the thirty-seventh. Who knows, the second half may appear wrapped around the body of another ancient Egyptian as unexpectedly as did the first.

But when words have been exhausted, when there is nothing further to say, or nothing further to communicate through written transmission, then the historian turns to excavation. The process of digging up the past is still a primary technique, and is still being refined. It is now being conducted with a camel's hair brush instead of a bulldozer, but we are the victims of bungling efforts in the past which have badly damaged information of incalculable value. For example, one of the early explorers of the Pyramid of the Sun at Teotihuacan near Mexico City decided hidden treasures were concealed inside the pyramid.[1] He began to dig, starting at the corner of this massive edifice. From his efforts, he learned nothing; but he permanently destroyed one corner of the pyramid which made it impossible forever to establish accurate measurements of the pyramid's base, thus eliminating the ability to learn precisely the mathematical formulations for the construction of the pyramid, and also the relationship of this remarkable building to the many other structures that are now being unearthed in this Mayan capital.

Because the dating of past events is often a key to understanding them, the science of dendrochronology gives us a reliable clue as to when certain events took place. In spite of the impressive term, it is merely the study of tree rings following methods proposed by Andrew Endicott Douglas, who was an astronomer.[2] In his study of the sun's spots, Douglas wondered if the cycles that he had observed for a period of time would be reflected in the rings of a tree, which as it grew, would reflect the meteorological conditions that were present during each year of its growth. He began his study with the bristlecone pine, the one tree that probably most happily confirmed his hypothesis, in a section of the country, namely, the southwest that has historically been subject to great variations in annual precipitation. A laboratory for the study of tree rings has been established at the University of Arizona, and by overlapping the bristlecone pine with the sequoia, it is now possible to push the dating process back to 4000 B.C. A fragment of wood which was used in construction, or was preserved without deterioration, can now be used to give an accurate date of activity on a particular site. The findings of this technique would be applicable only to the trees in one meteorological zone and would not necessarily be applicable to the ring sequence of a tree grown on the slopes of Norway, or on the seacoast of Morocco.

Another technique was developed by the American physicist Willard Libby, known as carbon-14 dating.[3] This technique depends on an analysis of the radioactive atoms of carbon 14 which are present in all organic substance. Carbon 14 deteriorates at a known rate. It is absorbed by all living things and, therefore, by analyzing bone, wood, plants, etc., it is possible to determine the date when carbon 14 was assimilated. This analysis is accurate back to approximately 40000 B.C. For some unknown reason, a fluctuation in the world carbon reservoir occurred in the past and readings older than 40000 B.C. are of uncertain validity.

The archaeohistorian may be at a loss as to where to start his excavation. In some cases, the object of curiosity is vividly evident to the human eye as the pyramids at Tiotihuacan, and there are hundreds of acres of still unexplored sites in the jungles of Yucatan. A new and promising instrument known as a magnetometer has proven useful as a prospecting instrument, and gives the curious person, when nothing is visible, some hint as to where he should begin to dig.[4] Because the structures of molecules and crystals in rocks are oriented to the magnetic pole, it is possible to detect large pieces of rock beneath the surface of the ground that have been placed in such a way that the magnetic currents of the stone are not in conformity with all of the surrounding rock. It is also possible to determine whether the earth has been disturbed in any significant fashion in the past.

"False color air photography" is one more device to reveal where the earth has been disturbed by the construction of canals, ditches, roads, or early settlements.[5] Through this type of photography, it is possible to locate ancient construction projects that are totally invisible to the person who simply walks across the sand. For example, scientists have been able to place the very extensive irrigation system built by the Incas of Peru, which brought into cultivation hundreds of thousands of acres of land along the Pacific coastline of South America, and which supported a population three times the size of the contemporary societies in this area.

The taking of blood samples requires enormous care, patience, time, accuracy and specialized technology. However, the information revealed has been informative. Apparently all native Indians from the tip of the Pacific northwest through South America to the Strait of Magellan have blood-type O. There are exceptions to be noted, but these exceptions are evidence of a racial mix. The number of people who migrated originally from Asia across the Bering Straits to Alaska were limited and may have consisted only of a few families. Curiously enough, this small group belonged to the O type blood family. A knowledge of this fact establishes that there were no successive waves of emigrants after them; and that the Eskimos found in northern Alaska and Canada are not Indians, but came from different racial stock.[6] The O factor is recessive, which means that it is carried on the X and Y chromosome, and any individual who is the product of a combination of parents, one of whom is O type and the other is either A or B type, will reflect the genetic characteristics of the A- or B-type person who is his parent. Other traits are sex-linked with the O blood type. One of the most obvious is beardlessness, which is a characteristic of all Indians in the Americas who have not been subject to crossbreeding. In his study of the Indians of Peru and of the population of Easter Island, which lies some thousand miles away from the South American shoreline in the western pacific, Thor Heyerdahl used the analysis of blood types to establish the line of migration to this island. He tested both the natives of Peru and the Polynesians who, he believes, crossed the enormous distances of the Pacific Ocean from west to east, using northern currents which swept them close to North America, and then back again to Easter Island.[7]

In addition to the A, B, and O blood factors, the Rh factor is also a clue; and as the science of hematology develops, additional techniques for determining the genetic characteristics of people may develop. The discovery of the double helix and its component parts in the DNA molecule opens up the possibility for tracing our genetic inheritance with a vastness that is bewildering.

Of all the amazing biological structures of the universe one that deserves our unlimited admiration is a single grain of pollen. It is so small as to be invisible. It is a protein which has a very hard shell and which can survive in some locations for hundreds and even thousands of years without breaking. Small enough to be unseen, tough enough to be indestructible, it is no wonder that with an exterior like a porcupine's it can enter the bloodstream through the mucous membrane and in some cases through the skin and induce highly irritating allergies.

Pollen grains at least 5,000 years old have been found mixed in the dust of the pyramids. In another context, a revealing discovery is reported by Magnus Magnusson in his book *Viking Expansion Westwards*. In a midden (refuse heap) of a Viking colony that lived in Greenland prior to 1300 was found a grain of corn pollen.[8] This tiny speck tells us chapters. Corn was not only the most important plant-breeding experiment of the Indians, it is also probably the greatest horticultural accomplishment of mankind. Corn, as we know it, was developed on the Mexican highlands in the Teotihuacan Valley. As everyone knows, the kernels remain attached to the cob and both cob and kernels are contained within a husk, and this means that corn is not self-seeding. Only by the intervention of man can the kernels of corn be removed and planted. Thus the ecology between corn and man is complete; and if man were not to intercede in the planting process, the production of corn would disappear from the earth entirely.

Corn became a staple of the diet of the Indians of both North and South America and was a gift of the New World to the Old. The presence of one grain of pollen in the midden heap of the Greenland Colony is evidence that they knew about corn and that it was part of their diet. This particular grain of pollen could have come from no place but North America. It could have been transmitted to Iceland in no way other than by the intervention of man, which indicates one of the first links of traffic from the United States eastward to Europe. If there were no regularized trade patterns between Iceland and the North American coast there was at least the minimal necessity for sufficient contact to transmit seed corn from hand to hand, as well as information as to how it should be planted and consumed. Thus we open a new chapter in the history of man, and the importance of man's use of food and other botanical materials which has resulted in the creation of the world's first Ethnobotanical Laboratory in the United States at the University of Michigan.

We have known for some time that the world's larder has been greatly improved, both in variety and quality, as a result of the contribution of the indigenous Americans. These people were involved in developing new food sources when the first European visitors came to these shores—perhaps as early as 2000 B.C. From the native Americans, the rest of the world received such gifts as corn, beans, squash, tomatoes, potatoes, rubber, tobacco, quinine, and sweet potatoes, to name a few. The state of nutrition of a culture has been studied and gives a key not only to the economy, to the standard of living, and to the commercial and trade activities, but also to the transmission of botanical information.

Some of the techniques used in ethnobotanical science are borrowed directly from the field of medicine. Foremost is the use of "coprolites" or petrified human excreta found in the oldest American settlements.[9] Through a flotation process, developed by medicine for the examination of tapeworms in human beings, it is possible, over the span of many days, for the coprolite to break down and rise to the surface so that the various elements may be analyzed. Substantial information has already been developed by the technique, but thus far it has not revealed to us why the Olmecs would suddenly desert one fully developed metropolitan center and move to establish another a great distance away; or why the population of Teotihuacan suddenly disappeared from the face of the globe leaving so many of its riches behind. The exhaustion of the soil by overcultivation of corn and other nutritional factors may have played a role and ethnobotanical research may eventually provide a key to this mystery.

Other promising research techniques include X-ray thermoluminescense.[10] Broken pieces of pottery are the universal thumbprints of the human race. When these pieces of fired pottery are reheated to a temperature equal to that at which the original pottery was fired, the material gives off a luminescence revealing the metal content in the clay that was used to make the pottery. According to *The New World*, this analysis of potsherds that were originally manufactured in 2000 B.C. in the Equadorian Andes revealed that pottery made in Tutishcainyo on the River Ucayali contained volcanic crystals of a recent eruption, and therefore the finished pottery was made of raw materials imported from a known distance. Studies made on pottery fired years later by the Incas showed further distances for the import of raw materials and an extension of trade deeper into the South American continent, and to the North American continent as well.[11]

The X-raying of mummies in the Cairo Museum by the College of Dentistry at the University of Michigan tells us more about the Egyptians than simply the degree of dental decay 5,000 years ago. Strange court secrets have been revealed by this process, and scientists have found evidence of early skull operations as well as cranial deformations that may have influenced the lives of the pharaohs.

The study of linguistics is only in its infancy. If Noam Chomsky is able to support his claim that human grammar is built into the nervous system, he may provide us with a new basis for the translation of languages, and also a method of tracing the shifts, changes, and origins of these languages. As a scientific Mr. Doolittle, he may be able to tell us things about race, class, and tradition that we would never suspect, either about ourselves or our ancestors.

Finally, there is experimentation, one of the techniques more difficult to devise and to use. Certainly the towering figure in this field is Thor Heyerdahl, who has demonstrated that it is possible to sail a raft built on the Inca model and cross the Pacific. He has also built a boat out of African totora reeds. He once very nearly made a crossing of the Atlantic, and on a second attempt—having learned something from the first—made the complete journey.[12] For some reason, there appears to be a conspiracy among professional archaeologists to ignore Thor Heyerdahl. The reason why is hard to understand.

5 • Utopian Mystery

WE have an inadequate picture of early societies in the Americas, for most either deceased before we found them or were shattered on their exposure to European civilization. What we know about the Indians of North and South America is biased, partial and inadequate. It is recorded from the European point of view, and the events that occur are not those seen by the eyes of the Indian.

There are exceptions, of course. Now that we have reduced the North American Indian to social impotency, completed our dispossession of his lands, and reduced the rights guaranteed him under the various treaties to virtual nonsubstance, we are finaly free to make an honest appraisal of our conquest of this continent. Thus, John Collier, former Commissioner of Indian Affairs, in his book *Indians of the Americas*, was willing to describe our degradation of the Indian.[1] Collier wrote with passion for greater understanding and protection of a minority. He appealed for honesty in the maintenance of pledges and promise. But he was not able to take that next step, and actually enter that ancient culture, to travel the byways of the Indian mind—the shadowland of wind, rains, Great Spirits, and the black-tailed deer. No one has every immersed himself enough inside that culture to write the history of the Treaty of Saginaw with the same sense of moral indignation he would use to punch out words to describe Hitler's absorption of the Sudentenland, or the German invasion of Poland during World War II.

Similarly, no one will ever be able to know the thoughts, the passions, the plans, and the dreams of the Aztec people. The destruction by Cortes of that culture and its people was so thorough that little information about the Aztec mind and spirit comes down to the present time. The Spanish Christians did their job well. Children of the Inquisition, they had a hardened intolerance that forced them violently to destroy everything they did not understand, or which failed to agree with what the Church had said. The callous destruction and awesome malediction done in the name of Christ still stupefies us with its terrible arrogance, and reminds us that some of the same spiritual spirochetes of a horrendous disease may be lurking in our veins today. Had the Spanish Christians taken time, and had they had the patience to learn, they might have found much in the religion and culture of these people to admire, and even cherish. But they did not have time. There was one thing they could understand and one thing they could cherish. That was gold.

There is one exception, however, to this vast expanse of ignorance, and that is our knowledge of the Inca civilization of South America. It was visited and studied by Europeans before its conquest, and a book describing the people, their way of life, and the organization of Inca society was published in 1516. This was almost twenty years before Pizarro, in 1532, with 150 brigands—and a few priests to bless their rape, mutilation, assault and robbery—were able to paralyze a nation of one million people, and do it so effectively that within a span of a few years that society was totally unworkable.

That book about Inca society was written by Sir Thomas More, the "Man for All Seasons"—a man of integrity and trust. The name of his book was *Utopia*, and it was printed in Europe—although not in England—twelve years before Pizarro's penetration of Peru.[2] In his book, which seems to be two essays of almost independent character joined together to make one book, More opens with a description of the English government, its violence, its inequities, its insecurities, and the defenselessness of those of low estate against crimes and misdemeanors as well as the venality of the ruling class. This doubtless is reason enough why More did not publish in England. The second portion deals with a society in which most of these failures have been rectified. More makes it clear that this society is no dream of his; rather, it is an account of a conversation he has had with a perceptive traveler to whom he gives the fictitious name of Raphael Hythloday.

In the opening narrative, the traveler describes what would clearly be an accurate account of a trip from the Atlantic coast of southern Brazil into the inner desert of Paraguay, and then on to the highest mountains of the Andes by means of what may have been one of the Inca highways. Finally, he arrives at a capital city located on a lake. On reaching this city, "the first vessels they saw were flat bottomed, the sails were made of reeds and wicker woven close together, and some were made of leather." This is a very close description of the unusual craft used then and now on Lake Titicaca and pictured in Thor Heyerdahl's book, *The RA Expeditions*.[3] These boats were woven of totora reeds which, according to Heyerdahl, had to be transported from Egypt and planted by hand in fresh water, because they grow not from seeds but from suckers put out along the root of the plant so that new shoots appear in shallow water from the mud below the level of the water. Heyerdahl had seen reeds like these before and boats woven together like this in pictures painted in tombs and caves in Egypt. It was by following the Egyptian pattern of early boat building that he was able to construct a craft that would sail across the Atlantic from Africa to South America, and then continue on to Peru where he found boats woven together in exactly the same pattern on a lake which was once the center of the Inca Empire. Thus it was Heyerdahl's assumption that long before the Inca Empire came into existence this same area had been visited by travelers from across the great sea who brought boat-building technology as well as cultural ideas with them, and who planted their ideas along with the totora reed in the memories of the Andean people.

While Hythloday managed to reach Peru at least thirty years before Pizarro, this description of Inca life may have been even older, and Hythloday may have been repeating tales of earlier adventurers to which he may or may not have added his own discoveries.

More wrote that Hythloday sailed on three of the four reputed voyages of Amerigo Vespucci, and jumped ship to make the journey across the continent in 1503. Scholars are reasonably clear that Amerigo Vespucci made at least three of the four voyages attributed to him, and the fact that More makes reference to a man for whom two continents are named, and whose achievements as an explorer have grown in stature with every century, had the effect of nailing his story to an acknowledged historical fact. There were, however, evidences of previous explorations from the east to the west.[4]

A map prepared by Andrea Bianco in 1448 portrays "an authentic island in distance 1500 miles to the west of the Cape Verde Islands."[5] The island which Bianco identifies on his map is a remarkable rendition of the Atlantic coastline of South America for a distance of 1,000 miles, including the Caribbean Sea. The actual distance from Cape Verde to Brazil is 1,520 miles,[6] which is a reflection of the precise navigational skills that mariners possessed in the century before Columbus. The fact of pre-Pizarroan voyages to Peru, however, is corroborated more conclusively in a careful comparison of *Utopia* with William H. Prescott's *Conquest of Peru*.[7]

Prescott was the scion of a wealthy family. As a student at Harvard College, he lost the sight of one eye and his vision in the other was impaired when a prankster threw a piece of bread at him across the dining hall. He became a scholar; and probably because he was largely dependent on others to read to him, he developed a memory of prodigious quality and was able to quote with perfection passages as long as fifty pages. He hired the most excellent scholars in Europe to research for him, and they were paid well to probe the Imperial Spanish files, look into private collections, and rummage through the archives of ancient bishoprics. It was not until Prescott's book appeared in 1847 that the parallels between More's *Utopia* and Incan society became apparent. Correspondences in More and Prescott not only in major outlines but also in seemingly trivial and unimportant details, revealed that two important books had been written about the Incan culture.

Peering through this small window in the unpenetrable wall of Incan history, we have a tantalizing view of life in that curious empire. What we see stimulates the imagination and arouses questions which are likely never to be answered. As indicated, the most sympathetic source of information is Hythloday, as reported by Sir Thomas More. Sir Thomas was sufficiently skeptical that such a society existed that he did not want his publication to be an offense to the authorities of England. The writings of More must be compared with that other source, *Conquest of Peru*. Prescott's scholarly agents worked intensively in the Royal Academy of History at Madrid. Their records were those of the victors and not of the vanquished. Prescott tried to be an impartial historian, but it is doubtful if his sources were able to yield more than a trace of the kind of objectivity for which he yearned.

It is said that if a man does you an injustice he never forgives you. For generations the Spanish Christians have had to justify the rape of a great civilization. They have had to make palatable the lying, deceit, and craftiness with which that annihilation was planned. Indicted by the finger of history, they have had to defend themselves against the charge of genocide—never surpassed until the rise of the present century—which effectively eliminated an entire race, innocent of any wrongdoing against the aggressor. The speed of that reduction from security to slavery was brutal and obscene. Pagan societies have been more compassionate. Our European ancestors could justify their acts only in two ways: One was to say that the Incan society was venal and cruel, and this justified the obliteration it received. The second was to say that the eradication of almost a million souls was necessary in order to bring the remnant to Christ.

What is reported here avoids the need to defend or accuse. It is a report on the Incas based on the scholarship of Prescott with an acknowledged indebtedness to Arthur Morgan, former president of Antioch College, who, working with the writings of Sir Thomas More, was able to bring us a creative synthesis. Additional research has been carried on constantly, including the brilliant discovery of Machu Picchu in 1911 by Hiram Bingham. In 1964 another American, Gene Savoy, using the same documents as Bingham, was convinced that Machu Picchu was not the city of last resort, and proceeded to find an even more remote Vilcabamba, a partial excavation of which has greatly enriched our understanding of Peruvian archaeology.[8] Such effort promises continuing surprises, although it is doubtful if any information will rise to cast new light on the dynamics of Inca society.

One could begin any place in describing this society, but certainly one significant fact is that it was a moneyless economy. Gold and silver were plenteous. These metals were used lavishly for artistic purposes, for personal adornment, for tableware and for decoration of buildings. One could claim the superfluity of gold made it unusable as a medium of exchange, but this is not accurate. It is always possible to establish an economy of scarcity through monopoly. The state could have claimed the gold—just as other nations have done in the past; and by keeping it sufficiently in control, the volume in circulation would not be excessive. Fundamentally, it was at this point that the Inca civilization diverged from the economies controlling us today. From Adam Smith to the International Monetary Fund, we have devised our economic mechanisms on the assumption of scarcity, and without scarcity a market cannot exist. In agreement with More, Prescott has this to say about the Incan economy of abundance: "Under this extraordinary policy, a people advanced in many social refinements, well-skilled in manufacture and agriculture, were unacquainted, as we have seen, with money. They had nothing that deserved to be called property."[9]

More described a society in which "they sow much more and breed more cattle than are necessary for their consumption, and they give that over-plus of which they make no use themselves to their neighbors. . . .there being no property among them."[10]

If money were not the token of value, what was it that replaced money as a symbol of worth?

The answer is work. From the Inca—the head of the state—downward, all participated in manual labor. In this democracy of work, the efforts of a manual laborer were considered as digified as the production of the royal family.

Emphasis was placed on high productivity, craftsmanship and a sense of individual responsibility for accomplishment. There were no graduations in the respectability of labor, with one exception: the slaughter house, which was unique in its repulsiveness. All forms of manual labor were considered equally dignified. No attempt was made to relate the quantity of reward with the status of the workman.

Every Incan was a farmer.[11] Whatever his primary craft or employment, he was required to give one-fourth of his working time to the exceedingly diverse agribusiness of the Empire. There were a variety of work schedules that could be elected. A person might work three months out of the year, or he might work one year out of four in agriculture. Normally he would go into the country where special housing and feeding facilities were available for those people who worked the countryside or the coastal areas. From the larger cities, groups of twenty would be assigned to work together, and to share the food and lodging that were part of the agricultural assignment. In the Inca society, some people preferred to farm the year round, and not to live in cities or villages at all. These people had the option to remain on the farm constantly, and such arrangements were not without advantages.

About a fourth of the Incan's working time was assigned to labor for the state. This was known as "tribute labor," and would reflect slightly less than the amount of income paid in taxes for government in the United States today. However, since the unit of value was work, this meant that in actual practice 25 percent of the individual's time would be assigned to the construction of roads, aqueducts, storehouses, palaces and bridges, or the repair of temples.

During the remaining half of his working time, a man might be employed as a craftsman or artisan, many of whom labored directly for the Inca and his family and the governor of the province. Production went into storehouses for distribution to the people.

The workday was standard throughout the Empire. It consisted of a six-hour day, normally three hours performed before dinner and three hours after.[12] The balance of an Incan's time could be devoted to his family, to the pursuit of his own particular craft, to the socializing with neighbors and friends, to the playing of games, to listening to poetry and the enjoyment of theater.[13]

As a laborer, the high commitment to the Peruvian work ethic evoked special praise from both More and Prescott. As an example, let us note what Prescott has to say:

> The face of the country would appear to be peculiarly unfavorable for the purposes both of agriculture and internal

communication. The sandy strip along the coast, where rain rarely falls, is fed only by a few scanty streams...the precipitous steeps of the Sierra with its splintered sides of porphyry and granite, and its higher regions wrapped in snows that never melt...might seem equally unpropitious to the labors of the husbandman.

The soil, it is true, was for the most part sandy and sterile; but many places...needed only to be properly irrigated to be susceptible to extraordinary production. To these spots, water was conveyed by means of canals and subterraneous aqueducts executed on a noble scale...Some of these aqueducts were of great length. One...measured between 400 and 500 miles in length...In this descent a passage was sometimes to be opened through rocks—and this without the aid of iron tools; impractical mountains were to be turned, rivers and marshes to be crossed, in short the same obstacles were to be encountered as in the construction of their mighty roads.

Many of the hills, though covered with a strong soil, were too precipitous to be tilled. These they cut into terraces, faced with rough stone, diminishing in regular gradation toward the summit; so that, while the lower strip, or *anden*, as it was called by the Spaniards, that belted around the base of the mountains, might comprehend hundreds of acres, the uppermost was only large enough to accommodate a few rows of Indian corn. With such patient toil, the Peruvians combated the formidable obstacles presented by the face of their country! Without the use of the tools or machinery familiar to the European, each individual could have done little, but acting in large masses, and under a common direction, they were enabled by indefatigable perseverance to achieve results and to have attempted that which have filled even the Europeans with dismay.

In the same spirit of economical husbandry which redeemed the rocky Sierra from the curse of sterility, they dug below the arid soil of the valleys and sought for a stratum where some natural moisture might be found. These excavations, called by the Spaniards "*hoyas*," or "pits," were made on a great scale, comprehending frequently more than an acre sunk to the depth of 15 or 20 feet, and fenced round within a wall of adobes, or bricks baked in the sun. The bottom of the excavation, a floor prepared by a rich manure of the sardines—a small fish obtained in vast quantitites along the coast—was planted with some kind of grain or vegetable. Under the patient and discriminating culture, every inch of good soil was tasked to its greatest power of production, while the most unpromising spots were compelled to contribute something to the subsistence of the people. Everywhere the land teemed with the evidence of agricultural wealth, from the smiling valleys along the coast to the terraced steeps of the Sierra which, rising into pyramids of verdure, glowed with all the splendors of tropical vegetation.[14]

As ecologists, the Peruvians receive high rating. They understood both the ecology and the economy of western South American better than we do today. Their long aqueducts, tunneled through the Andes Mountains, still remain. The irrigation canals carrying water for hundreds of miles along the western sea coast survive, and the shadowy outlines of their *hoyas* and local irrigation canals can still be traced. But this area has reverted to the vast and lifeless plains that existed before the Incan presence. Terraces built on the Andes, sometimes as high as 8,000 feet, were planted, cultivated, and harvested by hand. These still remain in spite of the fact that no maintenance work has been done on the terraced walls for 400 years. Virtually all of them are in good condition, although no crops grow there now. In the area where this imaginative use of land resources occurred, now live only a small fraction of a people who once flourished before the arrival of the Spaniards, and they produce only a portion of what the Incan was able to tease, wheedle, and force from this unpromising soil.

To organize the work force, the Incas divided society into ten groups or classes based on age. Whether Erik Erikson, the psychiatrist who separated "the ages of man" into eight categories, had ever seen the Incan plan of organization is unknown, but at least he makes no reference to it. The observations of the Swiss educator, Jean Piaget, would also be helpful in analyzing the Inca structure since he, too, deals with the problems of human development. However, long before these educators and psychiatrists had reached their formative conclusions, the Incan had worked out a structure of ten groupings based on an intuitive understanding of man's maturity and capacity.

The first four classes comprise those persons sixteen years of age or younger and probably include babyhood, childhood, pre-adolescence, and adolescence. No individual in this category was required to work. Between the years of sixteen and twenty, assignments were made to light manual work, such as the picking of coca and the harvesting of berries, nuts and other edibles of small weight and easy access. Between the age of twenty to twenty-five, young people were assigned to help their elders, assist with household chores, care for their older neighbors, etc. Between the ages of twenty-five and fifty, the Incan worked a full six-hour day, "paid tribute," which meant that he worked a fourth of his time for the state, and was normally head of a household. Between the ages of fifty and sixty, he was considered "half old," and was assigned to lighter work. After the age of sixty, no work was required.[15].

Buildings, dwelling places, and lands did not belong to individuals. They were produced by group effort and were considered possessions of the group. Cities and towns were constructed according to plan. In the center of smaller towns and villages, normally at the place where two crossroads met, were placed four storehouses in which the produce of the district, as well as other districts, was stored. Incans lived in apartment houses very much as city dwellers do today. The size of the apartment was assigned by the governors of the community based on the size of the family occupying it. A newly married couple would be given a smaller amount of space; as the number of children increased, the size of the apartment was augmented accordingly. As the children left the nest, the couple were expected to move into more modest accommodations. And of course, the payment of rent for these facilities was by the only medium of exchange known, namely, work.

The economy was based on "overproduction," and each worker was expected to produce beyond his own and his family needs. The four warehouses in the center of the city might be likened to a department store, or more accurately, to a shopping center, each storehouse specializing in a particular variety of goods. In this economy of abundance, each community was expected to accumulate inventory to meet its own needs for at least three years in advance with anticipation of drought, plague, or any other "act of God." In addition, it was expected to build reserves adequate to the needs of each of the adjoining communities for a period of three years.[16]

Each storehouse was a specialty center. One would provision corn, grain, beans, tubers, and other edible products of a staple nature. A second would provide meat, fish, fowl, etc. A third was stocked with furniture, cooking equipment, pots, bowls, including vessels of gold and inlaid work, and household utensils. A fourth would provide cloth and yarn, and other materials for clothing, weaving, etc. A citizen of the community would come to the storehouse and ask according to his need. A person might request more than he could consume, but evidences of this are so scarce as to be negligible. Hoarding was pointless in view of the fact that the excess could not be sold and taking more than one could consume was useless since one could expect to have his needs met as they arose.

One portion of the land—approximately a fourth—was assigned to the government and to the priests. The remainder was divided per capita on a lease basis, but the tenant had no authority to buy or sell. Leases were assigned for one year only.

In order to plan production and distribution, an exact census was carried out every year, recording each birth and death, and listing the number of people in each of the ten categories of economic activity.[17] The production of each province, city and town was also recorded—not only agricultural but the production of cloth, cotton, wool, and other household items. According to Prescott, no European nation had ever collected economic and production statistics so accurately or knew so precisely the inventories of the state as did the Incas of Peru. Until recently, it was assumed that the Incas prospered on the basis of a "closed economy" and planned and distributed only according to their internal needs. However, recent findings indicate otherwise. The thermal-luminescent analysis of pottery fragments in Peru indicates that some of it was imported from considereable distances, and that commerce may have traveled up and down the Pacific coast of South America, and possibly even to more accessible sections of western North America.

To sum up the economic side of Incan life, Arthur Morgan says, "In the production of consumer's goods and their distribution to give universal economic security, they succeeded more completely than any nation before or since."[18]

The weight of central government rested as lightly as possible on the shoulders of the diversified people who comprised this vast confederated Empire. The government chose the most promising young men in each community and each province for administrative and judicial responsibilities.

They were brought to Cuzco, the capital of the Empire, for training. When they returned to their hometown or province, they were given the title of "Inca-by-Privilege."[19]

The state placed two requirements on each citizen. The first insisted that he believe in one god, the Sun or "Inti," the only god, according to the Inca.[20] Although the imperial government was lenient in allowing each province and group to continue to follow its own special or peculiar religious beliefs and practices, it did require the construction of temples for the worship of Inti, and priests were trained and provided to lead the people in his worship. Inti was a god devoid of anger, vengence, pride, or a desire to punish offenders. In his religion there was no place for magic, incantations or auguries. It was a joyful religion that taught respect for all of Inti's children, even those unlike one's self. It rejected violence and upheld kindness and respect for life—even the life of animals.

The second requirement of the state was that everyone learn to speak the Inca language in addition to his local patois. Teachers were trained in Cuzco and perhaps in other centers, and were sent to the smallest villages in order to conduct classes. Thus, the Incans may have been among the first to establish a public school system. Religion and language unified the various nations and provinces inside the Empire, and the government provided equality of treatment, economic and personal security, and an economy of abundance.

Laws were few.[21] "With no private property to speak of, no disputes with citizens over business relations, no real estate to divide, or debt to collect, and with no offenses against property," the country could almost do without lawyers and judges. In judicial trials, all decisions were made within five days. In case of guilt, the sentences were severe. Some offenses were capital, but usually consisted of condemnation to slavery.

Few societies have ever achieved such economic equality, or have appealed so little to self-interest as the dynamic for economic action. This did not mean that there was no difference in rank in Inca society, but these distinctions were based largely on role in government. At the top was the Inca and his family. There were other officials vital to the functioning of the Empire, and each province or state had its Incas-by-Privilege. Observable distinctions in rank were not based on gold or medals, but rather on feathers. The use of woven feather stoles, capes, caps, and vestments were signs that elicited veneration on state occasions. The crown of the

Inca himself was nothing but a purple fringe around his forehead in which appeared two feathers of a rare and curious bird, the "coraquenque," found high in the rocky peaks.[22]

The word "Inca" carries three different meanings. First, this term identifies the supreme ruler of the Empire, a direct descendant from the ruling family. The first Inca was Monco Capac and the last was Atahualpa. Second, Incan refers to the Empire governed by the Inca. The boundaries of that territory continued to change, always enlarging, with a recent acquisition shortly before the arrival of Pizarro. Third, the word indicates the family of the Inca, including the descendants of the three brothers of Monco Capac and the four sisters who emerged from the cave at Paucartambo a few miles from Cuzco. The members of the Inca family were distinguished for having red hair, beards, a very white skin, and unusual height. They were known as the "Viracocha," which eventually became the generic term for white men. Pizarro when he first arrived was greeted as a "Viracocha," and in this century Heyerdahl on his visit to Lake Titicaca and Cuzco was immediately and unvaringly referred to as "Viracocha."

In comparison to the rest of the world, communications in the Empire were excellent if not unsurpassed. Two sets of roads, both more than 2,000 miles in length, were constructed on either sides of the Andes and through the Andes Mountains. One road system traveled along the plains and deserts between the mountains and the sea coast to the present town of Mendoza, Argentina, and from thence back across the Andes to Santiago, the present capital of Chile, and on to a terminus at Talca in the south. These roads, which are intact today, with the exception of portions across the coastal desert, are still considered engineering marvels 400 years after their original constructions. Arthur Morgan, himself an engineer, describes them this way:

The coastal highway was an enlarged part of an embankment raised above the sand, although in some stretches where drifting sand would have oblitered any highway, huge markers were erected to indicate the way. In the mountains, the road builders many times made great cuts through solid rock. They built high fills across ravines and swung suspension bridges across rivers and canyons.

A post house was located every five miles along these roads at which were stationed couriers who ran their designated five-mile stretch to the next post house with speed. In spite of the great altitude, it was possible for important messages to be transmitted along this highway at a rate of 100 to 150 miles a day. The Incas also constructed at appropriate intervals along the highways permanent barracks and military warehouses filled with ample food supplies. The capacity for quick troop movements to the frontiers, or to any place in the Empire, was thus a well-known fact. Freight was carried by llamas; and an unexpected need for supplies in one part of the Empire could be quickly remedied by moving food, equipment, personnel, and other commodities from an area in which there was a surplus to the locality in which there was a need.

Some curious blind spots inflicted the Incas. How could they have developed the enormous public highway system and built aqueducts 500 miles in length through the solid stone of the towering mountains and fail to discover the wheel and axle, or the Roman arch? They did come close to the arch with the trapezoidal doorway and the corbeled roof with construction techniques similar to those used by the Egyptians and Celts. Nevertheless, with the benefit of hindsight, it is hard to see how they could come so close to these discoveries and yet manage to overlook them.

In spite of the many creditable characteristics of Peruvian society, there were those aspects that were mordant and sordid. They were three in number: conscription, war, and slavery. All Incas were subject to military training. There was no standing army. The military exploits of the Empire depended on universal training and this in turn depended on all the population. The absence of a standing army may have had its benefits: this alone may have made it possible to achieve an economy of abundance. Furthermore, the Inca mind apparently appreciated the inconsistency of maintaining a standing officer corps with the otherwise equalitarian aspects of Inca society.

The absence of a professional military corps may have done much to preserve the personal freedoms that were unique to Inca society. In comparison with European nations of the fifteenth century, there was, with the possible exception of Switzerland, no country with religious freedom equal to that of Peru. The argument has consistently been made that a civilian army is preferable to a professional army if the objective of a society is freedom. But even at its best, conscription is always the invasion of one man's life as a precondition to the invasion of another's.

As for slavery, it consisted of two classes of offenders. The first were military personnel defeated in battle who refused the invitation of the victors to join the Inca Empire. The second were condemned criminals. It was possible for the second group to earn their way out of slavery. In certain cases, it may have been possible for the first group as well.

With their overpowering commitment to work, it was impossible for Incas to think of keeping a prisoner of war in a detention camp, nor was it possible to imagine criminals kept as semi-invalids at the cost of the state. Both groups of slaves were condemned to work in the slaughter house, which of all forms of work was the most repulsive. Nevertheless, the meat had to be slaughtered and the slaves were given the job. Military slaves were apparently given the job of cleaning up after sacrifices performed in the temple. Any kind of killing was objectionable to an Incan. They banned the killing of animals for sport. Hunting of all sorts was illegal. But in spite of their inherent distaste for bloodshed, they continued to press outward the borders of the Inca Empire, both to the north and to the south. This paradoxical attitude deserves further comment. Let us hear first from More:

> They would be both troubled and ashamed of a bloody victory over their enemies, and think it would be as foolish to purchase as to buy the most valuable goods at too high a rate. And in no victory do they glory so much as that which gained by dexterity and good conduct, without bloodshed.
> They very much approve of this way of corrupting their enemies by persuading them to quarrel among each other, although it appears to others to be base and cruel; but they look on it as a wise course to make an end of what would otherwise be a long war, without so much as hazarding one battle to decide it.[23]

In much the same way Prescott observes:

> Their policy toward the conquered forms a contrast no less striking than that pursued by the Aztecs. The Mexican vassals were ground by excessive imposts and military conscriptions. No regard was had to their welfare and the only limit to oppression was their power of endurance. They were made to feel every hour that they were not part and parcel of the Nation, but held in subjugation as a conquered people. The Incas, on the other hand, admitted their new subjects at once to all the rights enjoyed by the rest of the community, and though they made them conform to the established laws and usages of the Empire, they watched over their personal security and comfort with a sort of parental solicitude. The motley population thus bound together by common interests was animated by a common feeling of loyalty, which gave greater strength and stability to the Empire as it became more and more widely extended. The policy of the two nations displayed the principle of fear as contrasted with the principle of love.[24]

The Inca Empire, it should be known, began as a very small community, namely, as a single city in which a small group of people, the Viracochas, through intellectual ability and strength of character achieved their leadership. According to legend, the Viracochas, or the Incas, "walked out of a cave." That is, they suddenly appeared from nowhere. They gradually extended the border of their city-state, largely through maneuver, persuasion, inveiglement, diplomacy, and genuine offers of productive cooperation. While battles must have occurred, they were few and enemies must have been baffled by the fact that they had nothing to fear by becoming victims. In fact, the process of losing was often the passageway to partnership in a more secure, more prosperous, and in some respects a freer citizenship than they had known before. With such a reputation preceding them, it is possible to understand how the people of the west coast of South America—who had had a long history of intercommunication and common government in the past—found it was impossible to struggle against the Inca, and in this they were totally unlike the enemies of the Aztec.

The ruling family remained small, possibly no more than a few hundred. To this group must be attributed the constellation of ideas that gave birth, life and unity to the Inca Empire. Did this group bring with them the technology, the engineering science, the skill of masonry and construction, the knowledge of ecology, the vast understanding of irrigation and cultivation, the craft of government, the social skill of unifying people of dissimilar background and temperament, the ability to cultivate loyalty and commitment, and the intellectual power to conceptualize an economic system unlike any the world has ever known? Did these innovations come from other sources, other climates, other traditions, other cultures? Or were these ideas all potential in South American society but never fully exploited at any previous time? Thus far, we have no certain way to know.

From the smallest beginning, the Incas extended themselves to one of the greatest empires the world has known in the span of only one hundred years, and this with a minimum of bloodshed. The Inca power continued to grow gradually with important innovations during the next century. And suddenly it came to a close.

The Viracochas were still a small family; they knew the value of cohesiveness. They managed to dominate by the quality of their thought. They generated a society which many of us today would say is "contrary to human nature." At the same time, they solved problems that are still high on the agenda of mankind as he approaches the end of the twentieth century.

But who were the Viracochas? Who were the Incas?

We can only surmise. But whoever they were, they were not Indians. They were white. They had red hair and red beards. They were tall. Pizarro made special note of their skin color, saying that they were even whiter than the Spaniards. A study of mummies[25] and the burial remains of members of the ruling family confirms that their hair color was auburn. When subjected to microscopic analysis, the hair is round and glossy and unlike that of any other indigenous Indian in either North or South America. A measurement of the skeletons reveals that they were several inches taller— both men and women—than the other peoples of the west coast of South America. A study of the physiognomy of the skulls[26] of these people shows a facial bone structure that would given an appearance scarecely different from North Europeans. From the known physical characteristics, scientists continue to ponder, "Who are the ancestors of these people?" With the rapid advances in biology and anthropology, their surmises become increasingly sophisticated, and someday the answer to this 400-year-old question may be found. As for their descendants, it is hard to say whether there were any. If there were, they must have been few, although for a few years Pizarro continued to prop up a puppet Inca as a device for confusing those tribes and colonies who were unable to understand that the structure of government had been totally destroyed.

The events in that destruction went like this: Pizarro set sail from Panama and headed south, skirting the sandy wastes of the Sechura Desert but with the towering Andes always in sight to eliminate the need for a helmsman, a star, or a compass to keep him on his course. Along the way, the conquerors would stop at occasional villages which were unusually friendly and courteous from which they gathered valuable information. They learned of the Incas' magnificent highway system; the riches of gold and silver that were not only the possessions of the Inca, Atahualpa himself, but the royal family, the provincial governors, and even the common people of the land.[27]

As Pizarro continued his journey down the coast, he saw the magnificent canals and waterways that had been contructed to make the barren plains fertile and was impressed with the quantity and the quality of the produce, and the evident health and vigor of the people. At each place he gathered further information on the vast riches of the Empire, its organization, and the Court of Atahualpa and his family. After acquiring a sufficient supply of gold and silver ornaments, as well as rich tapestries, he returned to Panama to prove his claims that great wealth lay in store for a conquering army. He left two of his soldiers behind in the city of Tumbes. Not only had his reconnaissance of the coastal area been extremely valuable in terms of the intelligence he was able to bring back in preparing for the next stage of his operation, but the great courtesy, kindness and generosity of the friendly Indian made it possible for him to set up "centers of influence" he could use for future conquest. He presented himself as a Viracocha, or "Child of the Sun," and plotted his own carefully concealed attack on the Empire of the Inca. The Indians identified him as a member of the royal family, and therefore a person in whom they put their trust.

In January 1531, Pizarro assembled for himself in Panama a modest little army of approximately 150 men. A few soldiers were armed with crossbows and harquebuses. Twenty-seven cavalrymen with war horses were included in the invading group. In addition he had two falconets, light mobile cannons which may have had more impact in creating psychological havoc than in destroying the enemy. Before departing, Pizarro and his little army attended a mass in Panama's cathedral, where their flags, their weapons, and all the men were consecrated to a "Crusade against the infidel." There is no doubt that Pizarro was convinced that "God was on his side, and every soldier under his command knew that he went forth a Christian Soldier to wage warfare in behalf of the gentle Jesus."

It was the intention of Pizarro to strike his first treacherous blow at the city of Tumbes, where he had been received as a friend, hosted as "one of the children of light," and from which fellowship he had cynically departed with pledges of love and friendship. However, the wind and ocean current blew him off his course and he was forced to land on an unknown shore. He proceeded inward to the province of Coaque, where the Spanish Christians fell on the unsuspecting natives with "sword in hand," and from whom, according to Pizarro's own account, they won for themselves a rich "store of gold and precious stones." From this unexpected source they seized approximately 20,000 *pesos de oro*, which at the present value of gold would be well over $200,000; this was sent back to Panama. After the town was thoroughly plundered, robbed, and raped, Pizarro continued on to the island of Puna, from which he planned an attack on his old friends at Tumbes. The attack came, and the fall of Tumbes was the most peaceful ever. Recognizing in Pizarro an old friend, who came back asking that as a representative of the pope and Charles V of Spain, the natives submit to his authority, and the Indians—knowing not the least of what he was talking about—willingly complied. They entertained him royally; and as he set out for the foot of the Andes, Indians everywhere provided him with food, offered him friendship, and gave him shelter at night. Wherever he went, he noted the high prosperity of these Indians in a land smiling with flowers and rich with orchards. He envied the prosperous fields laced with an elaborate irrigation system, the source of which he could only guess.

The report of these events were swiftly borne to Atahualpa along the magnificent Inca highway system. There is no doubt that he must have been troubled, puzzled, and alarmed. The little band of Spaniards approached the Andes on the well-constructed highway, and were almost overwhelmed by the magnitude of the task that lay ahead of them. About Atahualpa's attitude or military resources they knew little. But they could see the towering peaks of the mountains; and for a people who lived at sea level, the thought of scaling mountains to a height of more than 16,000 feet made them wonder whether after having completed their journey they would have at the end of it the energy necessary to carry out the military operations that would be involved. Intrepid warriors they were, and they began to climb the mountains, achieved the pass, and descended the Andes on the opposite side to the town of Cajamarca. As they approached, the town was strangely quiet. They discovered that at Atahualpa's orders the town had been abandoned, fully stocked and supplied, the entire population having retired across the river where they were encamped with the legions of the Inca who had traveled thus far to meet these other "Children of the Sun."[28]

After stationing his men in the well-provided city, Pizarro was the first to cross the river to establish personal contact with the Inca and his Court. It was the practice that when the Inca traveled, his family, consisting of several hundred persons, traveled with him. In addition, there was an array of almost 2,000 Indian officials and soldiers camped on the surrounding hillsides. Unknown to Pizarro, an army of 35,000 had been garrisoned several miles down the highway under the command of the brilliant General Challcochima.

Pizarro approached the Emperor with deference. With all the modesty at his command, he presented himself as a gracious person, full of admiration for the Inca's magnificent achievements, his courage, and his imperial majesty. There was an exchange of tokens of affection; and before departure, Pizarro extended an invitation to the Inca and his entire Court to come across the river the following day to Cajamarca for dinner and for brotherhood with other "Children of the Sun." The mission had been a success.

Splashing across the river, Pizarro hurried to meet with the other members of his command to lay out the plans for the following day. He carefully studied the city to determine where the cavalry should be placed to block quickly an exit for the visitors. Through some of the small streets that approached the central square, they prepared hiding places for members of the infantry. They concealed their two cannon near the front of the plaza area. Infantrymen were placed out of sight on the top of houses, in towers, and in the temples.

Apparently Atahualpa had some misgiving about this invitation because the next day he proceeded to the bank of the river and then halted, saying that he would attend the repast on the following day. Pizarro sent urgent messengers saying that the food was ready and begged the noble Inca to come at once. For whatever reason, this second invitation allayed Atahualpa's hesitation and he went forward on his litter, carried by members of the royal family.

Pizarro gave each of his soldiers detailed instructions on the action he planned for the Inca's visit. They were to remain totally and completely out of sight. They would begin the reception when Pizarro fired a shot and held high in one hand a white handkerchief. His soldiers had nothing to fear. He reminded them that they had come this great distance for the purpose of converting souls to Christianity. The Church had blessed their crusade. God had blessed their crusade.[29] They were held in the loving arms of Jesus and nothing to fear from pagan infidel. If God was for them, who could be against them? Pizarro was right! No Christian was killed; no one was seriously injured.

Atahualpa moved into the courtyard on a litter. He was surrounded by a group of two hundred Viracocha and approximately two thousand Indians of the highest rank who were "Incas-by-Privilege," heads of colonies, provinces, cities, and holders of other high office.

As this procession came to a halt, there emerged from the surrounding buildings two shadowy figures. One was the Dominican friar, Vincete de Valverde, and the other was his interpreter. Valverde explained to Atahualpa that he and the other Spaniards had come a great distance to teach the Inca people the true faith, and he proceeded to give a brief analysis of Christianity and its principal dogma. He concluded his homily by saying that the pope was God's agent on earth and he had given Charles V, King of Spain, responsbility for conquering and converting the Inca Empire. He implored Atahualpa to renounce his god and turn to the one true God and become the servant of Charles V.

It is possible that Atahualpa grasped only dimly the first part of this discourse; however, he had clarity about the concluding remarks. He responded with great force and emphasis, saying "Your emperor may be a great prince, and I am willing to hold him as a brother. As for the pope of whom you speak, he must be mad to talk of giving away countries that do not belong to him." The translation of this conversation may be imperfect. The only record we have of what happened that day was written by the victors, and certainly in terms of the vile deeds that were about to occur they had reason to weigh every syllable to make it justify Pizarro's claim that the Christians were the agents of God and had come to save souls. Nevertheless, in spite of all the temptations to misconstrue and misremember, there is a majesty in the words of Atahualpa that even four hundred years have not been able to dim. What he said in effect was, If you have come to teach us, that is one thing; but if you have come to control us, that is another. Obviously the intention on the part of the Spaniards to confuse two missions angered him, but Atahualpa's statement was one of conciliation. "Your king may be a great prince and I am willing to hold him as a brother." It was a strange juxtaposition of ideas in a moment before every man in the courtyard was to lose his life that the mention of brotherhood came from the pagan and not from the priest.

In those few moments before he launched one of the most vicious mass murders in human history, the heart of the Spaniard may have been momentarily warmed by the utterance of that word, "brother." If this is the case, he must have quickly brushed the thought away. He had Christ's work to do.

The moment Pizarro had long awaited had arrived. He waved high a white handkerchief and fired the shot that was a signal for the atrocity to begin. The cannon thundered at point-blank range. Men rose from the rooftops and volleyed into the crowd as rapidly as they could reload their weapons. The cavalry charged into the crowd with sabers flashing in the sun to cut people down like a field of ripened grain. All the members of the Viracocha stood solidly around Atahualpa without moving, without flinching, without flicking a lash. In a half hour, there lay dead 2,000 Indians of the highest rank, including the "Incas-by-Privilege," governors, administrators. Also dead were 200 members of the royal family and the Inca Court. Only one man was spared: Atahualpa himself.

That day the Inca Empire came to an end.

The process of dying and decomposition continued for another ten years. Pizarro took Atahualpa into custody, and promised him that if he could fill a room as high as the Inca could reach with gold and golden objects, he would be given his freedom. A treaty was drawn up which both the Inca and the illiterate Pizarro signed. The orders were given and the kingdom was scoured from one end to the other, and the room finally filled with gold to the height agreed. With this mission completed, Pizarro summoned the Inca to a trial on trumped-up charges, order him garroted and his body burned at the stake.

Pizarro then created puppet Incas, presumably to win the love and loyalty of the Indians, but none of these plans worked more than briefly. The Spanish soldiers on looting expeditions fanned out to the tip of the southernmost portion of the Empire, and any semblance of personal leadership or control on the part of the Inca or members of his administration was quickly and permanently destroyed. In less than ten years, chaos had descended as the Spaniards began to fight among themselves. Pizarro murdered his most able captains and lieutenants; and finally, falling victim to a cunning plot on the part of his subordinants, Pizarro himself was killed.

Some of the family of Atahualpa and a few of the Viracocha may have survived. In later years, there were those who came forward, claiming to be descendants of the royal family. Women who made such claims were occasionally married to officers or ambitious Spaniards striving to become Peruvian nobles. Quite aside from the validity of their ancestral claims was the pitiless epilogue to the fall of the Empire which began as the last curtain went down and the exploitation of the Incas by the Spaniards began with murderous savagery. Every able person was forced to work in the mines as many hours as he could survive, day and night, digging out more gold, more silver, more copper, more gems. In the process of working the mines, they drafted those men who knew how to make the irrigation system work; they drafted the men who were able to plan the planting and cultivation of crops; they called up the persons who organized the rotation of agricultural workers from cities to farms. The result was famine, and the consequence of famine was plague, and plague brought death. Whole cities and communities, mostly of children, the aged and women, were stricken down overnight. The vast granaries which stretched for hundreds of miles along the Peruvian coast were returned to a desert. Finally, even the supply of Indians to work the mines was reduced to a trickle. Many of them doubtless prayed that they might have died with Atahualpa in the square at Cajamarca.

Why could an empire of more than a million people fall to a force of 150 soldiers in such a brief period of time? Part of the explanation lies in the fact that the Inca state was highly centralized so that when the Inca, the Viracocha, and the "Incas-by-Privilege" were removed, the organization of the society fell into complete disarray. But there is another explanation of considerable importance: the Spanish soldiers with greater stature, white faces, and beards appeared to be members of the Inca family. They were immediately identified by the Indians as Viracocha. Toward persons of this appearance, the Indians had developed an overpowering sense of respect. With this reverence, bordering on worship, it was not easy to disregard the Spaniards, much less destroy them.

And thus we return to the question, "Who were the Viracocha?"

The first of them arrived "out of a cave"—presumably from nowhere. There were eight—four men and four women. Everything we know about them to date confirms our belief that they were European. Their place and time of landing on the South American continent we do not know. If they brought a written lanaguage, which seems unlikely, they did not share it. They carved no hieroglyphs, raised no steles, and kept their records on quipu, or knotted string. This method of storing information was of Egyptian origin, and as a method of keeping records it was exceedingly accurate, and permitted the development of a form of mathematics whose secret has been lost. In spite of the use of quipu, the Incas probably were not of Egyptian origin. There was little else in the cultural mix that they brought to this great Inca Empire that is reminiscent of the culture of Egypt—with the possible exception of Ikhnaton, the religious heretic whose teachings could be harmonized with the Inca way of life. However, the prospect of an Egyptian origin for the eight original Viracocha seems too remote to deserve further attention at this time.

They could have been Phoenicians, or Phoenician priests, or they could have come from the largest Phoenician colony of Carthage, or even a lesser colony of Marsailles or Cadiz. However, this means that their crossing would have to have been made no later than A.D. 300 and it seems unlikely that even though they were part of an original colony of great numbers, they would manage to survive wandering through the South American wilderness for eight hundred years before coming to the place of maximum influence at Cuzco. Furthermore, the Semitic features of the Phoenicians, or the dark complexions of the Carthaginians and the residents of other Mediterranean countries hardly suggest the physical type of the Viracocha for which we have adequate anthropometric evidence.

The organization of the Inca society bears a strong intimation of Christian values, although there are Christian scholars who would dispute this point. However, considering that the Peruvian empire began to take shape in the early 1300's, some of the economic, religious, and personal values do not appear out of harmony with gentler Christian brotherhoods of the time, and with some of their monasteries.[30]

The fair complexion and the red hair raises the question as to whether they might have been Irish monks. The objection will be raised that monks are sworn to celibacy. Therefore, it would have been impossible for four red-haired, fair-skinned Irishmen and four fair Irish women to "come out of the cave" and proceed to propagate and carry on this lineage for two hundred years. However, it was not until the middle of the twelfth century that the Roman Church officially required celibacy of its priests and clergy. Futhermore, these redheaded men and women might have been Celtic Christians, and it is known that the Celtic Church was influenced by Christianity in the East. Even today in the Eastern Orthodox churches clergy are permitted to marry. It is only the bishop who is denied a wife. No, the religious inhibitions alone would not have prohibited Christian Celts, even priests, from entering matrimony.

There is one difficulty with this thesis, however, and that is that the ocean currents from Ireland do not easily lead to South America, and if such a colony of pilgrims did set out for "The Island of the Blest," they would have undertaken a very long journey to South America with the likelihood of being swept off their course by contrary ocean flow.

There is one further possibility: Fair-complexioned, redheaded Christian pilgrims also lived in Spain. They were Celtic Christians, too, and they must have known that navigation from Spain and Portugal had traditionally followed the ocean currents south to Africa, and thence across the Atlantic to the Caribbean or the east shore of South America. The art of shipbuilding, we know, had reached a level where such a voyage had been possible for many years, even centuries. Having been subject to persecution, they may have been the dreamers who came to the American shore led by a hope to establish a Christian community according to their best lights and according to the best resources available to build it.

Any of the above could have happened. But let us part with the Inca and his Viracocha here, and leave an unfinished chapter with additional pages still to be written.

6 • Utopia: Land of Persons or Prisoners

MAN is an animal who can conceive of an "utopia," and nowhere has this impulse been more evident than among those who settled the Americas. Indeed, the search for "utopia" may have been the hunger that led to the discovery of these shores.

The very word "utopia" is partisan. On the one hand, it is an epithet—a term of derision. The "utopian" is the impractical, the useless, the visionary, the dreamlike, the unrealistic. But at the same time, utopian thinking also implies a criticism of the status quo. In some societies, the description of utopia has been the one legitimate means of pointing out the failures of an existing state of affairs.

Sir Thomas More, like Isaiah, knew that his portrayal of a good society would inevitably be regarded as a thinly veiled criticism of his own country—and therefore the job of being a utopian was a hazardous occupation: one to be undertaken at a safe distance and at considerable personal risk. Curiously and paradoxically, utopian thought has been held to be at the same time both irrelevant and dangerous. As a consequence, the history of utopian thinking is of enormous politcal significance, because utopian thought with the promise it holds for a better life has not only been a critique of the present but has been the warehouse of significant ideas for our own political future.

From the earliest days of human history, the lands to the

West have been associated with utopian thought. Ulysses dreamed of "The Happy Isles." Brendan the navigator sailed off to America—the "Islands of the Blessed." The Mayflower Compact was written in the cabin of a storm-tossed ship on its way to Plymouth and can be read today as nothing but a utopian document. The charter of Pennsylvania was borrowed directly from More, from James Harrington's *Oceana*, and from some of the less practical beatitudes of Jesus. Rhode Island with unprecedented religious freedom was a utopian community, and the people of Massachusetts are today the beneficiary of utopian dreams. The Constitution of the United States, if it had not worked, would have been labeled as a utopian document, and in parts of this world today it is still regarded as unreal in its utopianness.

Furthermore, the United States became a magnet that drew persons from around the world, but particularly from Europe to the land where the "good life" could survive without persecution. Thus the Quakers, Mennonites, and Anabaptist Brethren were among the first to come. Later other commuinities flourished: the Shakers, the Mormons, the Amish, the New Jerusalem, the New Economy, the Bruderhoff, all left an indelible trace on the pages of American history. Perhaps most of the emigrants who came to these shores arrived in search of a better life and thus "the American Dream" becamse a diffuse and undefined utopia to which we all bear some unconfessed allegiance.

Like other settlers, the pilgrims who moved to Peru were propelled by a powerful hunger for justice and good will. In this they were no different from others who made that same fateful trip across a wide ocean, generally never to return, in search of a future denied to them at home. They appeared in Peru suddenly. "They came out of a cave"—four men and four women, all with red hair—the cave doubtless where they rested the night before they were discovered by native Indians. This was the beginning of Peru, and what was to become the Inca world.

To deny this legend is possible, as well as all that flows from it: that these eight strangers settled among them and provided the intellectual leadership for a small city to grow into one of the great empires of history; that they were able to ignite the fire for works of engineering that takes one's breath away, and more particularly to design an economic system without a market; that they could be audacious enough to conceive a society without a working class—to do all this and found it on an altruism which—Peru excepted—we know is unworkable.

Many scholars claim that this was not altruism at all. Peru was, they say, the first totalitarian state. The occupation of each person was chosen for him. His daily task was based on a plan drawn by the central bureaucracy for the construction of highways, the building of terraces for farming, and the erection of apartment houses. The Peruvian could not pass from one city to another without a permit, and in most cases his marriage was arranged for him. His was a life without freedom. And while the achievements may have been brilliant, they were the achievements of the state and not of the individual.[1]

One of the acid tests of every utopia is "Who does the dirty work?" In Peru, the answer was "Everyone." For the Inca, the backbreaking toil was farming on which everyone depended, and in Peru everyone was a farmer. As indicated earlier, the civil service of the capital city was organized to spend three months out of every year assigned to farming, or one year in four working in agriculture. A comparable arrangement would be for the bureaucracy in Washington, including Congress and the Supreme Court, to spend three months of every year manning the assembly lines at General Motors. Unthinkable! But the arrangement had one profound advantage—it avoided the identification of a "working class." Or, more precisely, under this dispensation, all people belonged to the working class.

At least one exception was made to the rule of dirty work, and that was the slaughter house. So fundamental was the work ethic that it would be unthinkable to the Incas that offenders should be excused from labor by confining them to jail or secreting them behind stone walls. To be sure, the prisoner performed the vilest possible tasks, but his prison was to wear a badge without which he must never appear, and this emblem designated him as an offender. The badge was his incarceration.

To say that the Inca had no freedom in his work assignment is an overstatement. Apparently he did have some occasion to choose the work schedule that suited him best, some option as to where he was assigned, and some choice with whom he worked. If he was healthy and within the proper age bracket he did not have the freedom of electing whether to work or not to work, but he had the freedom of requesting that he be transferred to another province. The Incas noted that those who requested the largest number of transfers were generally the least productive workers. If he was bright and energetic he also had the freedom to accept training for the civil service. The first two ranks in public service were filled on an elective basis. From that point on, greater responsbility was awarded on merit, and the highest rank of all "Inca-by-Privilege" was a combination of service and the equivalent of "life-peerage."

Because of commitment, community pressure, and national planning, Inca men between twenty-five and fifty years of age were required to work without exception and apparently they did work like demons. It is because of this characteristic that Inca society has been most subject to criticism, and described "as freedomless as an anthill." In this connection, it would be interesting to compare the work commitment of the Inca with that of the serfs and peasants in Europe during the same span of the fourteenth and fifteenth Centuries, or even with the freedom of the assembly-line workers at the Ford Motor Company today.

The Inca was required to work at times on projects that were stupefying in the amount of energy required for such tasks as terracing whole mountainsides for thousands of miles in the Andes. The workday was, nevertheless, six hours. Probably no society, before or since, has allowed its people so much time to follow their own pursuits. The Inca became skilled at handicrafts, particularly pottery and weaving; he invested time in the government of his community; he played games, attended the theater, and listened to minstrelsy; he spent time in conversation with his friends and in family pursuits. But all of his free time was not invested in high-minded ventures. In classic simplicity, Prescott described the greatest enjoyment of the Inca in these words: "Dancing and drinking were the favorite pastimes of the Peruvians."[2]

The great triumphs of the Incas, however, were not their highways or their terraces, their aqueducts or the palaces, their agriculture or their communications, but rather were in the realm of personality; in the state of affairs they achieved within themselves, and in their relations with their neighbors.

"In Peru. . .the mutual helpfulness of small community groups was not abandoned in the process of empire building, but was carried over and was made the dominant characteristic of the structure of states and of a great empire."[3] When this pattern was broken down by the Spanish conqueror, there followed a deterioration of technical competence, a decline in resources, a loss of population, and a vast demeaning of the life of the people.

As a policy of state, all effort was made to strengthen the myriad of small communities with their traditional virtues of neighborliness, cooperativeness, affection and good will. And without doubt, these were the pervasive qualities which Prescott sensed in all he read about Peruvian life, and which made him say that the Inca people would have been wholly predisposed to Christianity if only the Spanish captains had come in search of God rather than of gold.

Among the nations of the world, the Peruvians have been unique in upholding the quality of life as the most significant goal of government, and it is for this reason that the utopia of Peru, as well as all utopias, may speak to the condition of man.

As we become aware of the great cost of stress in our own society, we find awakened interest in a society that reduced stress to its lowest threshold. While the Inca society was competitive in recognizing service to society, it was noncompetitive in all of the other interpersonal relationships on which the quality of life are built. The Peruvian economy demonstrated that it was possible to have hard work done and that the motivation for such achievement could come from some other fires than the desire for personal gain.

Stress, we now know, reduces our resistance to disease and weakens the immunization systems of the human body. Most specifically the increase of stress is seen in a growth of heart trouble, stroke, and possibly cancer. So clearly is stress and physical disability related to each other that, according to Arnold Toeffler, predictability is possible, and a simple check of those persons who have undergone great emotional erosion becomes a dependable guide to morbidity and mortality as well.[4] Much stress in our society could be eliminated. In the case of stress that is unavoidable, the crippling effects on the individual could be greatly modified if the stricken person were to find himself in a supportive community during these periods of trial.

So the Peruvian society was great not bcause of the magnificent structures it erected, but because of the microstructures which were effective in nurturing and sustaining the individual, that which supported him in the crises of life, and in the changes that were as inevitable then as now. This has always been the focus of utopian thought— the fate of the person. And because every intelligent and active person is to some degree utopian, the life in ancient Peru is not without relevance today.

7 • The Skill of the Phoenicians

THE shipping of the entire known world was dominated by the Phoenicians for a thousand years from 1200 to 200 B.C. Paul the Apostle booked passage on Phoenician vessels known as the "Ships of Tarshish," and it was Jonah's trip on a Phoenician ship which terminated in his encounter with a whale.

Unrivaled in their mastery of shipbuilding, seamanship, and business administration, the Phoenicians established a monopoly of ocean trade which was unequaled, even in the Mediterranean, until Venice began to dominate trade and demonstrated the advantages of new management techniques such as double-entry bookkeeping. Ships of other nations did venture into the Atlantic well before the time of Christ and by careful, coastwise navigation they did manage to travel as far North as the Rhine and as far south as Morocco.[1]

The Phoenicians were of Semitic extraction, and rather than give their attention to land warfare they poured their energies into the exploration of the sea. Why this kind of romance—one might almost call it a passion—for the sea arose among these people is now difficult to trace or understand. However, their ability as mariners is mentioned as their most singular and important characteristic, dating from the earliest reliable historical record. They were recorded in the ship list of Homer, who speaks with admiration and awe of the "Black Ships" of the Phoenicians. Although the events in the *Odyssey* occurred about 1300 B.C., and oral sagas began to develop shortly after the fall of Troy, Homer lived about 800 B.C. and may have actually reflected a later-day admiration for Phoenician seamanship.

The Phoenician ships were able to travel about twice the speed of other vessels. They were able to traverse directly across the Mediterranean, whereas the ships of other navies preferred to sail close to the shore and normally camped at night on land. The largest of the Phoenician ships were also about three times the size of the other vessels that plied the Mediterranean and about three times larger than the vessels that Columbus used to make his first Atlantic crossing. The ancients were awed and amazed by Phoenician seamanship. Homer says, "The ships are steered by thought," and he explains further, "for there exist no pilots among the Phoenicians, and there are no rudders at all such as other ships have. But the ships themselves know the intentions and the minds of men. They know where to go and how."[2]

The Phoenician sailors themselves were of great courage, and Homer says that among them there was "no fear." They sailed through fog and hazardous conditions that other navies would never attempt. Their Black Ships were considered "unsinkable," and the mariners were reputed to have some kinship with the gods who favored them—all except Poseidon, the god of the sea, who, because he was uanble to sink their ships, "bore a great grudge against them."[3]

Dominating the sea lanes of the Mediterranean, they became great colonists. Queen Dido of Phoenicia founded the city of Carthage, which became the trading capital of the Mediterranean and grew to a population of more than a million. Their dominanace put them in obvious competition with Rome, and in the course of three Punic wars they slugged it out with the Romans. The first two wars were indecisive, but the Carthaginians lost the third, and the city was wiped out forever. Had they confined their warfare to the waves, they might have been invincible, for it was only on land that the solid, stolid, Roman legions were able to grind them to

Phoenician ships not only dominated the Mediterranean, they also operated on established commercial lines as far north as Sweden on the Gulf of Bothnia, and as far south as Dakar and Sierra Leone on the Atlantic coast of Africa. Over a span of several centuries, repeated colonizing expeditions were launched into the Atlantic, and under the captaincies of Himlico and Hanno a single flotilla may have contained as many as 30,000 pilgrims and some may have been even larger. Phoenician shipbuilders were hired to construct vessels for other navies, and when their coastal cities were under foreign domination, were impressed to build navies for their conquerors. After Alexander the Great wept because he had no more worlds to conquer, someone whispered in his ear about distant lands beyond the seas. He therewith commissioned the Phoenician shipmasters to build the greatest navy ever. Some of the vessels were veritable behemoths designed to carry as many as 500 passengers and crew—a size not to be duplicated again until the eighteenth century. To what destiny this navy sailed is one of the mysteries of history.

bits. With the fall of Carthage, the Phoenicians' control of the Mediterranean was ended; but they had also founded other colonies of great importance, including the present city of Marseilles in France and the Spanish port of Cadiz.

The Phoenicians discovered and colonized the Canary Islands. By combining Phoenician sea power with Egyptian science and mathematics, one of the greatest voyages of discovery in human history was carried out by Phoenician ships. The Pharaoh Necho understood very well that the earth was round and that the sun was the center of the solar system almost 2,200 years before the idea was rediscovered by Copernicus. He had also calculated with reasonable accuracy the approximate circumference of the African continent; and in order to demonstrate his point, the Pharaoh on the throne of Egypt from 609 to 595 B.C. hired Phoenician ships to circumnavigate Africa from east to west. While Necho's scientists had estimated the travel distance to be 9,000 miles, it proved to be considerably farther than that because of the fragmentary information they had on prevailing winds and ocean currents. As a result, the Phoenician flotilla ran out of food and they had to stop along the way to grow a crop of wheat in order to continue the journey. Success crowned the voyage, however; and three years later they jubilantly met the Pharaoh at the delta of the Nile.[4] Herodotus, who is the historical reporter of this exciting expedition, tended to discount the story because the sailors reported that after their ships had entered the Atlantic Ocean the sun began to rise on the starboard side of the ship rather than on the port. Herodotus believed the sailors had become seriously confused, but later historians agree that the fact that the Phoenician sailors stuck to their story that the sun began to rise on the opposite side of the ship only gives credence to the accuracy of the report that the voyage occurred. Obviously, if the Phoenician sailors under Egyptian tutelage were able to make a voyage in excess of 9,000 miles and return safely, these same seamen were capable of crossing the Atlantic. At its closest, South America is only 1,520 miles from the Coast of Africa, and it would be possible virtually to float across this expanse, as Thor Hyerdahl describes this voyage in *The RA Expedition*.[5]

There are conflicting claims as to who discovered the Azores. Samuel Eliot Morison discounts the report that a cache of golden coins minted by the Carthaginians was later discovered on the islands. However, it is interesting to note that from the Island of Corvo in the Azores, it is only 1,054 miles to Cape Race, Newfoundland; and as the mariners probed westward and found one island after another, there is every reason to believe that these early seamen would continue to risk sailing west in search of further islands, which would bring them at last to North America.[6]

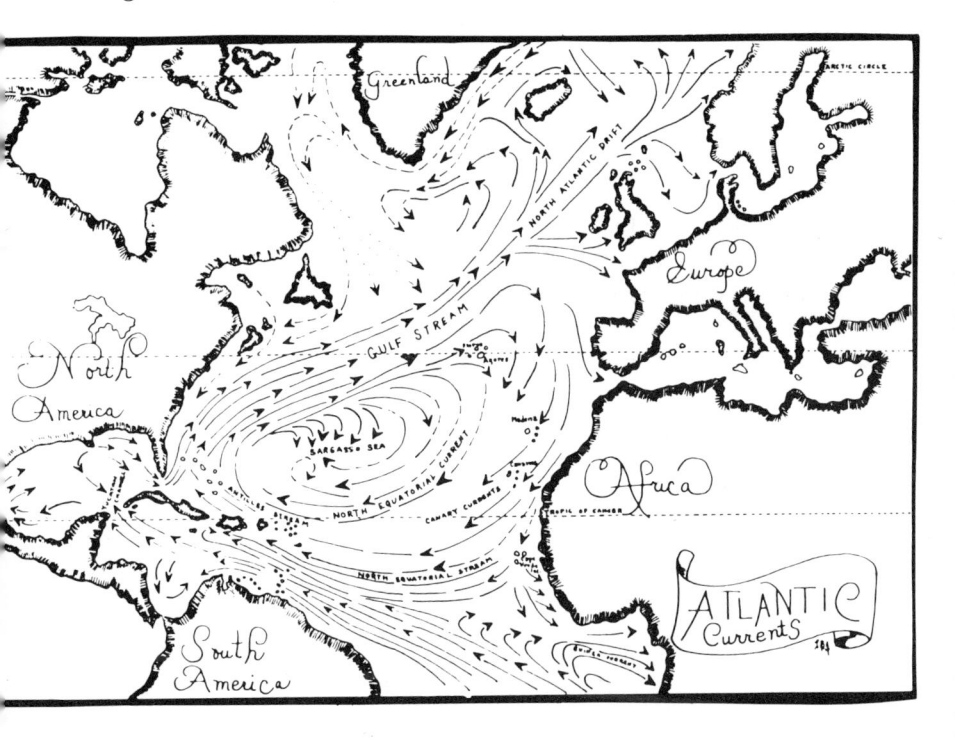

Carthage was at the time the largest city in the world, and was a center for industry, manufacturing, commerce, and exchange. Like all competitive industrialists, the Phoenicians were protective of their sources of supply and their markets, as were the great Portuguese mariners who made navigational history 2,000 years later. It was the Phoenicians who introduced amber into the channels of world commerce. Noted for its unique electrical properties, amber became an item of high demand, whose source was unknown except to the sailors who brought it into port. We now know that this amber came from the Baltic Sea, near the modern Polish city of Gdansk. Amber is also found in spots along the Swedish coastline in the Gulf of Bothnia, and occasionally along the English coast. Knowledge of navigational hazards, ocean current, and wind characteristics were carefully guarded secrets, because on the basis of this knowledge the Phoenician monopolies were protected. Their boats brought to Carthage enormous quantities of murex, used for dyeing the royal purple. But more important, the Phoenicians developed a virtual monopoly of tin, which is the one metal essential for making bronze out of copper. To preserve secrecy, the mining was actually done by the Phoenicians themselves, and one colony was founded on the northwest coast of Spain, and another on the tiny Isle of Scilly, not far from Lands End, at the tip of Cornwall. In view of their monopoly of this metal, which is not found in the United States, it was very likely Phoenician ships brought tin for the use of the early Indian moundbuilders in this country,[7] who combined it with native copper to make bronze implements found in Indian burial mounds from 200 B.C. to A.D. 200; or, what is more probable, Phoenician traders exchanged bronze implements made in Spain, along with purple dye and glass beads for the furs and leather goods of the Indians.

The legend of America existed in Europe for a millennium before Christ. Hesiod, a Greek poet and historian of the eighth century, and Homer both refer to "the Happy Isles" which lay somewhere beyond the Straits of Gibraltar. These Elysian Fields were the final abode of the good. The memory of this happy land has long been part of the Greek inheritance, and the English poet, Alfred Lord Tennyson, in his poem *Ulysses* describes the great warrior-sailor setting out on a new voyage:

> For my purpose holds
> To sail beyond the sunset, and the baths
> Of all the western stars, until I die.
> It may be we shall touch the Happy Isles,
> And see the great Achilles, whom we knew.

The poet who described the "Happy Isles" was certainly not confusing his dreams with the Canary Islands, which had already been discovered by the Phoenicians and which decidedly did not have the salubrious climate attributed to it by Homer or Hesiod. Off the coast of Africa, the Canary Islands have a climate not unlike Egypt's except for humidity. The mountains are craggy and volcanic and unlikely to be confused with Elysium.

The "Happy Isles," and the "Isles of the Blessed" mentioned by St. Brendan, the Irish navigator, may have been the same land of which Homer sung. Certainly the Carthaginians, who were now being faced with an encroaching Sahara Desert and a diminishing of the north African breadbasket, were interested in colonizing rather than further expanding the metropolis of Carthage, which was becoming ever more dependent on shipping for its survival as a habitable center of culture and commerce.

The Carthaginians had already gone beyond the Straits of Gibraltar to create colonies in northern Africa and on the Atlantic side of the Iberian Peninsula. In one account, the Carthaginian sailor, Captain Hanno, was placed in command of an expedition consisting of sixty ships having fifty oars each, chosen to transport a colonizing party of 30,000 men and women with adequate supplies to travel 3,500 miles beyond Gibraltar into the Atlantic.[9] Whether the colony under the captaincy of Hanno made it to the Caribbean is unlikely but is not recorded, although the dates of this colonizing expedition do synchronize with the emergence of culture on the peninsula of Yucatan. It is likely that at least a portion of these colonists were carried to other locations that seemed to be propitious for the start of a new city, a new life, and a new land.

In 1975, stones were found on the Atlantic coast inscribed in a southern Iberian alphabet and attributed to Hanno, dated from 480 to 475 B. C.[10] Debate about the authenticity of these inscriptions has arisen, and there will be no discovery, now or ever, that will not disturb someone's view of what was, or what should have been history.

Among those who discount the Phoenician discovery of America is Samuel Eliot Morison, whose opinions deserve careful attention.[11] In addition to the stones found on the Atlantic coast in 1975, a large number of Phoenician descriptions have also been found in Brazil; and Bernardo da Silva Ramos claims to have catalogued 2,800 of them. Morison dismisses these inscriptions on the basis that the Carthaginian language had a very simple written style somewhat comparable to the ogam writing of the celts, and what appears to be an inscription may actually consist of scratches or natural formations in the rock. Other scholars agree with Morison. But in 1968, Professor Cyrus H. Gordon of Brandeis University, after giving the matter cool attention, has announced that he believes the inscriptions are authentic.[12]

Assuming the genuineness of these inscriptions, it would appear possible that if the ships of Captain Hanno's fleet reached the shores of Mexico, a decision was made to divide the colonists into three groups with one settling in Mexico, one going south to present-day Brazil, and the third sailing north for a landing on the coast of North America.[13]

8 • The Wisdom of the Egyptians

THE weakness of the Phoenician was his failure to develop the intellectual capital on which further advancement in in trade and shipping were dependent. Therefore, an effective working relationship with an intellectual center, such as Egypt, became a highly useful arrangement, and thus the wisdom of Egypt supplemented the skills of Phoenicia.

With the exception of the Delta, Egypt is a country a thousand miles long and ten miles wide, at least in terms of the distribution of population. All traffic moves up and down the Nile. Northbound traffic can float with the current down the Nile, with sea anchors, rudders and oars, to keep the boat in the main current. In Egypt the wind blows from north to south and therefore a square-rigged ship is able to sail upstream from north to south. A greater speed can be achieved by the addition of oarsmen, and most military vessels were equipped to take advantage of both breeze and brawn. The Egyptian contribution to the maritime arts was great, but was largely confined to river traffic.

Egyptian shipbuilding evolved over a span of 3,000 years from the construction of woven reed boats, comparable to those used by Thor Hyerdahl in his crossing of the Atlantic.[1] Later ships were constructed of wood imported for this purpose, and were variously designed to serve as utilitarian freighters, passenger ships of various class, military vessels, and pleasure craft. Magnificent and highly decorative craft were developed specifically for the Pharaoh, and carried the body of the king to his final resting place. However, these boats had to be towed by others largely for parade purposes, and thus failed to contribute to the development of the maritime sciences.

In the entire course of Egyptian history, one memorable expedition was that of the Pharaoh Hatshepsut. The Emperor was a woman, but she was not a queen—she was a Pharaoh, and she wore an artificial beard on ceremonial occasions, and the sculpture shows the beard neatly hooked over her ears. Her great expedition is magnificently detailed on the Pharaoh's burial chamber.[2] On the basis of this recently reconstructed burial temple, it has been possible for Bjorn Landstrom to work out the accurate measurements and construction specifications of these boats for his book, *Ships of the Pharaohs*.[3] The flotilla consisted of five magnificently designed and decorated ships which were constructed on the Nile and sailed down to Coptos where they might have been disassembled, but very likely were pulled intact across the desert and placed in the salt water of the Red Sea. Because of the character of their construction, these ships were of the coast-hugging type, but they continued their voyage successfully to the land of Punt, the present-day Ethiopia. The expedition was a success, although the boats may never have been out of sight of land. Laden with tropical fruits and animals, they returned to the Port of Quseir, where again the the ships were pulled across the desert to the Nile. This exploration, which occurred in approximately 1400 B.C., was the high-water mark of Egyptian naval penetration, although Thor Heyerdahl has conclusively demonstrated that it was possible for ships of earlier design, namely, those made of totora reeds, to cross the Atlantic, and that the early boats were more seaworthy than the colorful boats of Hatshepsut which were built primarily for river traffic. If reed boats were used to cross the Atlantic—and Heyerdahl believes they were because of the evidence found at Lake Titicaca, in Peru, as well as on the west coast of Mexico and on Easter Island—no official records have come to light to indicate these voyages were sponsored by the government or recorded in any official document. However, the Egyptians were eager to explore the world not only for conquest and imperial pride, but because their astronomy had taught them more about the earth than any European knew until the time of Copernicus. Essentially, the Egyptians could supply the information the Phoenicians needed if their capacity as shipbuilders and sailors were to reach its maximum potential. The joining of Egyptian knowledge and Phoenician skill was like all good marriages— a partnership that benefited both people, and in this case that benefit was substantial.

For an understanding of the state of Egyptian astronomy and mathematics, consider the observations of Peter Tompkins in *Secrets of the Great Pyramids*.[4] Tompkins may be one of the greatest living expert on Egyptian mathematics and astronomy. He has devoted an appreciable portion of his life to the further study of Egyptian specialties, such as surveying, geography, and history. Tompkins summarizes his knowledge in these words:

> Whoever built the Great Pyramid, it is now quite clear, knew the precise circumference of the planet, and the length of the year to several decimals—data which were not re-discovered until the seventeenth century. Its architects may well have known the mean length of the earth's orbit around the sun, the specific density of the planet, the 26,000 years' cycle of the equinoxes, the acceleration of gravity and the speed of light.[5]

If, indeed, it is true that the Egyptians knew that the world was spherical and knew its dimensions, they possessed data which was not retrieved until two hundred years after the voyage of Columbus, as well as other scientific information which was not recovered until this century. And if they did accurately possess this and other celestial information to which Tompkins makes reference, they were adequately equipped to navigate accurately from Egypt to America and plot their way home again.

Parenthetically, credit for first calculating correctly the diameter of the globe goes to Eratosthenes, librarian of Alexandria in the third century B.C. He was an Egyptian but he was fortunate enough to have his works translated into Greek, which gave them a currency the older Egyptian works were not to achieve until Egyptian hieroglyphics were mastered in the latter half of the nineteenth century. Although Eratosthenes' conclusions were correct, he did make a number of errors in his calculations. Fortunately, these errors canceled each other out so that the final figure is remarkably close to the one that we now know is scientifically correct. It is now also clear that Eratosthenes was merely citing older Egyptian information about the circumference of the earth without really understanding it. Eratosthenes did carry out an experiment by calculating the slant of the sun's rays in a well at Alexandria at the moment of the summer solstice when the sun was known to be directly overhead, based on observations in a well at Aswan in the Tropic of Cancer. What he did not know, and what his predecessors of 2,000 years before had correctly calculated, was the calendar for the tipping or declination of the earth on its axis. Incidentally, the declination of the earth on a 26,000-year cycle which Eratosthenes did not know, but which the earlier Egyptians did, was rediscovered by the Mayans in Mexico in about 200 B.C.

According to Tompkins, the surveying which went into establishing the baseline of the Egyptian pyramids as well as other dimensions were more accurate than those attained for several centuries after Columbus. The north-south axis of the pyramid is due north with an error of less than one-fourth of an inch in a distance of 375 feet—so precise that compasses are calibrated by the pyramid rather than the reverse. We now believe that neither the Great Pyramid of Cheops nor any of the other pyramids were constructed as funerary memorials to kings or noblemen. They were all constructed as astronomical observatories, although the early step-pyramids of Saquara were originally built on top of mastabas, which were burial mounds where the royal or noble dead had been previously interred.

The period of pyramid building in Egypt was remarkably brief. The first efforts at the construction of a pyramid occurred about 2900 B.C., and the pyramid-building era of the Egyptian culture was complete by 2700 B.C. During this span of two hundred years, the astronomers, architects, stone cutters and masons were experimenting with the kind of structure, the appropriate slopes, the necessary dimensions of the base, the unit of measurement, the length of the star observation shaft, and the siting of the structure itself. As far as the astronomers were concerned, this required several centuries of patiently recorded observation before all of these calculations were accurately made and the information properly assembled. When the architects and designers understood fully the failure and even the hair-breadth inaccuracies of all of the previous experiments, the Pyramid of Cheops was constructed, and although it no longer has the golden cap it once had and the fine-grained limestone surfacing has been stripped from its side, it nevertheless was the ultimate and final achievement of Egyptian astronomy, architecture, surveying and mathematics. Cheops was the climax. The job was done; there was no need to repeat it. After its completion, the age of pyramid building in Egypt came to a close.[6]

Efforts to penetrate the Great Pyramid of Cheops probably were made soon after its construction. These efforts have continued down to the present day, and even with the recent use of the most sophisticated X-ray technology it is not possible to say whether the internal contents of this great mass of masonry—exceeded in volume only by the Boulder Dam—has been satisfactorily completed. However, in about A. D. 813, a young Arab king by the name of Abdulla Al Mamum hired a small army of stone cutters and managed to force his way into what is now known as "The King's Chamber." He found only one object—and it is clear that this chamber had never previously been opened after it was originally sealed. The object was a lidless coffer cut from a solid piece of chocolate granite containing granules of feldspar, quartz and mica.[7] There is no stone like it in Egypt, nor for that matter is there any evidence that comparable stone can be found in Africa. Myth has it that the stone came from Atlantis or from America. About the stone in Atlantis we can only speculate, but comparable stone can be found in America.

Some fanciful ideas about the pyramids have been slow to disappear. The idea that they were massive tombs constructed for Egyptian kings was dissipated by the fact that the ancient workmen built more pyramids than there were Pharaohs. As the inadequacy of each structure became apparent, they redesigned the basic plan with revised construction.

The unpleasant impression that the pyramids were constructed with slave labor has had a more tenacious life, but an article in the *Scientific American* rejects this idea and explains the organization of the work forces.[8] Some two thousand men worked year-round cutting the stone on the east bank of the Nile. The great mass of manpower worked only during the flood stage of the Nile, which was a period of enforced idleness when no one could till the land, but when it was possible to ferry stone from one side of the Nile to the other. The construction of the pyramids were more a public works project which had the political and spiritual purpose of providing a useful instrument that would benefit all Egypt. It would thus help in unifying the Upper Kingdom and the Lower Kingdom, giving Egypt a new cohesiveness based on shared work, common interests, and united loyalties. Egyptians were involved in an undertaking that had human, moral, and religious significance. Those who labored knew they were building for all time. In being part of that which was immortal, they too became immortal. How many of us have the sense of being involved in an undertaking of permanent grandeur? Perhaps this sense of participation was part of their remuneration. On the back side of the 250 million stones—and of course we have seen the backs of only a few—are found messages of encouragement, rivalry, jest, and homely humor—hardly an evidence of slavery and depression.

Reference has already been made to the chartering by the Pharaoh Necho of Phoenician ships to circumnavigate Africa about 600 B.C. Certainly this was a voyage of discovery that would appeal to any curious sailor of the time; and from the standpoint of the Pharaoh, the demonstration that Africa was a continent and therefore accessible by ship had more implications for the Egyptian Empire than building another wing on the Temple of Luxor. We wish we had more detail on the day-to-day progress of this voyage, and as more papyri become available, it is entirely possible that we will learn. The voyage started at the Delta. Phoenician ships must have sailed up the Nile to the City of Coptos and then were pulled across the Wadi Hammadi to the Red Sea. This may seem like an almost impossible human exertion; but when one remembers that during the days when the Apostle Paul visited the city of Corinth, ships were pulled across the Corinthian Isthmus by hand rather than sail around the Peloponnesus. When one considers how high Corinth is above sea level and how far those ships had to travel by land, and that this was a normal process on a traditional trade route that made Corinth a port city, though it was separated by a considerable distance from the sea on either side, the pulling of Phoenician ships across the Wadi Hammadi would have been a relatively modest undertaking. Compare it also to the later achievement of Cortes, who built a flotilla of forty boats on the Atlantic and hauled them more than 200 miles to the lake surrounding the Aztec city of Tenochtitlan, crossing two mountain ranges of 10,000 feet in the process. Thus, moving boats across dry land to put them back into the water was well understood by mariners. The Egyptian navy under Hatshepsut 800 years before had accomplished almost the same feat, and doubtless less historic ventures had repeated a similar transnavigation of the land many times in the interim between Hatshepsut and Necho.

It is entirely possible that Egyptians were present, even in considerable numbers, in the colonizational voyages of Captain Hanno in 400 B.C. also described in the previous chapter. However, in terms of the impact on North America, the clearest and most extensive evidence is found in the Mayan civilization of Mexico. Thor Heyerdahl has explored and explained the evidence of Egyptian influence in South America, and through secondary transmission to Easter Island. Elsewhere in North America, stone fragments and inscriptions have been found that hint at an Egyptian presence in North America. But one of the most fascinating cultural evidences are the Micmac Indians of North America.

In 1851, Henry R. Schoolcraft, Commissioner of Indian Affairs, reported to the U. S. Congress that after having made a thorough survey of the American Indians, it was necessary to report that "reading and writing were altogether unknown to them."[9] Naturally, this did not include the Cherokees, who in effect adopted the English language, in addition to their own spoken dialect. Neither did it include the Micmac Indians who, although related to the Algonquins, lived in Acadia, Canada, and thus were not studied by Schoolcraft in preparing his 1851 report.[10]

A French priest by the name of Pierre Maillard translated into the Micmac language a Mass, Scripture, selected psalms, and other liturgical writing. Some of these writings are interspersed with Latin directives or equivalences. Sent to Acadia when Cardinal Richelieu dispatched the first French missionaries to work among Canadian Indians, Father Maillard became the author of a 450-page book in Micmac hieroglyphs, which also included a catechism, a church history, and a translation of hymns. This book was printed in Vienna from original documents contributed by Father Maillard at the time of his death in 1762. A copy of this remarkable book is in the Widener Library at Harvard University, and a considerable number of additional copies are found in European libraries.[11]

The Micmac hieroglyphs are remarkably similar, and in some cases identical with Egyptian hieroglyphs. Furthermore, the Micmac meaning of these glyphs is in almost all cases identical to the classical Egyptian meaning. But old beliefs die hard, and scholars suggest that the Egyptian hieroglyphs were taught to the Micmac Indians by French priests who believed that it was easier to teach them hieroglyphs—specifically the Egyptian hieroglyphs—than to introduce them to the Latin language. These scholars claim that Maillard used hieroglyphs as a method of teaching Christianity because of the straightforward character of Egyptian grammar and syntax, and because of the visual aid to learning that the hieroglyphs represent.

But there is no possibility that Maillard first learned Egyptian, then translated his Latin into Egyptian and taught the Micmacs with the second language. Fortunately, the dating is clear. The 450-page book, printed in Vienna in Micmac hieroglyphics, is dated 1762, or sixty-one years before Champollion received Napoleon's prize for translating the Rosetta Stone, as well as the inscriptions on the Cleopatra obelisk which Napoleon brought to Paris from his Egyptian conquest. The possibility that Maillard might secretly have cracked the code of the Egyptian language sixty years before Champollion appears to be untenable for a number of reasons. It disregards the statements of the other French priests, including Father Sebastien Rasles, whose missionary work began in 1690 and who became familiar with the Micmac system of hieroglyphic writing when he first contacted them at the beginning of his missionary enterprise. Also, in 1738, more than a hundred years before Champollion, Father Maillard prepared a small grammar and dictionary, entitled *Hieroglyphic Manual of Micmac*. Barry Fell in *America B.C.* reproduces page after page of Micmac writing, both drawn from the Vienna publication and also from manuscripts prepared by Fr. Maillard in the course of his lifetime of experience with this Indian group. The Micmac hieroglyphs reveal that the Indians had a considerable knowledge of metallurgy; and interestingly enough, their symbols for silver and gold are almost identical to those used by the Egyptians.

If, then, on the basis of dating, it is evident that the Micmac Indians were not taught hieroglyphs by the Europeans, the question arises, who did teach them this form of written language? The answer has to be pressed back to at least A.D. 200, because at that time the ability to translate the hieroglyphs was utterly and completely lost and not rediscovered until the time of Champollion. This means that whoever the teachers were, whether colonists, visitors, merchants, missionaries, traders, they would probably have to arrive on the American shores sometime before the advent of Christ, or shortly thereafter. It is possible they might have been members of Hanno's Phoenician colony.

Other provocative specimens of what appears to be Egyptian writing are found in Davenport, Iowa, and in the Indian mounds of West Virginia.[12] However, dating problems still exist; and furthermore, these finds are tantalizing fragments in comparison to the thorough knowledge of the Micmac language derived from texts published in the 1700s as well as

The Wisdom of the Egyptians • 127

from the voluminous personal papers left by Fr. Maillard himself.

The spoken form of the Algonquin language shows the absorption of other linguistic influences, including that of the Celts; but the written form of the language, although showing evolution over the span of two thousand years or more is remarkably faithful to the original Egyptian.

The old portraits of the Algonquin Indians of the East coast pictured them as closely resembling the Europeans, particularly the southern Europeans of the Mediterranean type. When occasionally they are portrayed in European dress, it is difficult to separate them from European colonists. It is, therefore, quite natural that integration took place, and somewhat more rapidly than with the Indians to the West with stronger Mongolian features. As a result, only a small portion of the Algonquin people occupy their tribal lands, although a recent decision of the U.S. Supreme Court (1976) has altered the landholding statutes of the Indians in Maine based on a treaty signed between the Algonquin nation and the original European colonists. As a consequence of these new developments, we may now see some resurgence of earlier cultural forms.

An engraved stone known as the "Davenport Calendar" was found in an Indian burial ground in Iowa in 1974, 1,500 miles away from the Micmacs. This stele, or marker, was immediately branded a meaningless forgery. While dating problems do exist, it appears to be one of the few trilingual texts in existence, and is written in Egyptian, Punic, and Iberian languages. The text has to do with raising a monument at sunrise on the first day of the year. The marker is now in the Putnam Museum in Davenport, Iowa. The Egyptian writing certainly antedates Champollion, but no one really knows how this curious stone found its resting place in an Indian burial ground.

What the Micmacs and the Algonquins generally have to say about their origin is worth listening to. The earliest report seems to have been made in a letter by John Johnson, written in 1819, and later printed in *Archaeologicae*, who said, "The people of this nation have a tradition that their ancestors crossed the sea. They are the only tribe with which I am acquainted, which admits to a foreign origin. Until lately [i.e., 1819], they kept yearly sacrifices for their safe arrival in this country. From where they came or at what period of time they arrived in America, they do not know."

One other coincidence deserves to be noted. In the Micmac language, there is a hieroglyph that can be found in any standard Egyptian dictionary and is defined as "the checkerboard of Aton." The word "Aton" became a symbol for the religious revolution led by Ikhnaton at the end of the 18th Dynasty, or approximately 1320 B.C. The Micmac word for god "Atnaquna" is remarkably close to "Aton," the monotheistic god of Ikhnaton, and also is reminiscent of the name of god in other societies where non-Indian influences have been present.[13]

The Inca empire and roads

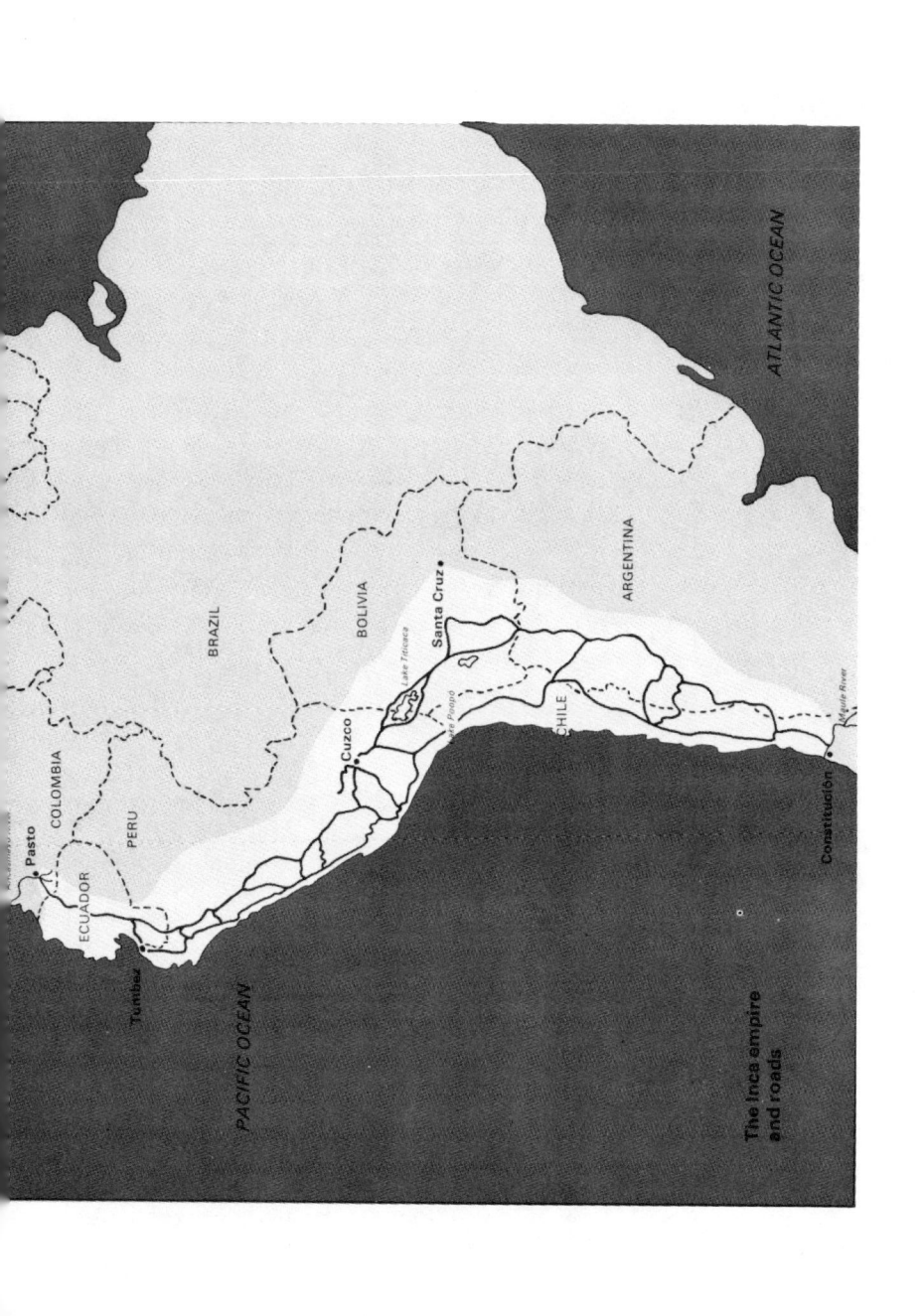

The roads built by the Incas before Columbus are still in use today and are regarded as engineering masterpieces.

Boats on Lake Titicaca are woven from totora reeds transplanted from Africa. The process of constructing the boat is accurately described in Egyptian texts and pictures.

The City of Teotihuacan was laid out on a grid plan which required precision in surveying. The unit of measurement used for the construction of the city is identical to the Egyptian cubit which checks perfectly with all calculations including square root.

9 • The Mayan Synthesis

CIVILIZATION in Mexico continued on an even road of evolution from approximately 2000 B.C. to the present with two exceptions. The first occurred about 600 B.C. with the introduction of a new set of cultural, linguistic and moral ideas that had the effect basically of enriching the indigenous Indian culture without destroying it. The second major cataclysm occurred in 1519 when that long, continuous growth came to a sudden halt with the invasion of Cortes, who not only annihilated all of the existing social forms but caused the population of the nation to drop to five percent. With the repopulation of Mexico, the Spanish, Christian, European culture became the dominating influence; and whatever Indian influences continued to survive did so because they were capable of assimilation into the new culture brought by conquistadores.[1]

There were approximately four eras of Mayan civilization prior to Cortes. All of them grew in the central area of Mexico, the Valley of Oaxaca, the Yucatan Peninsula, or in the area that is now Mexico City and its environs. These four eras were the Olmec, the Teotihuacano, the Toltec, and the Aztec. Each had its characteristics, but each also carried over a great deal of the past. In some respects, each one of these cultures overlaps the others, but there were regional variations within each of the four groups, and a tracing of these lines would involve us in detail that is not now directly pertinent to the impact of European and Mediterranean civilization on the peoples of Central America.

The Olmec tradition matured from 900 to 300 B.C., and other than regional variations there were many facets that suggest overseas influence that may have been African or Egyptian in origin. Included in this list are the Step Pyramids which are similar to those of the Zoyzer Pyramids in Egypt as well as those found during the rule of other kings of the Nile valley.[2] Hieroglyphic writing, both on stone and on sheets of parchment very much like books, were produced until the conquest. One of the first students of Mayan hieroglyphics was Augustus Le Plongeon, who prepared a table comparing Mayan hieroglyphics with Egyptian. After having devoted most of his life to this effort, he came to the conclusion that approximately one-third of the ancient Olmec symbols had the same meaning in the Egyptian language, and that the grammatical forms of the two languages were similar. An analysis of his findings is found in Peter Tompkin's book, *Mysteries of the Mexican Pyramids*.[3]

Monumental stone sculptures appeared in the Olmec heartland with no known antecedents, and at the same time craftsmen began to create small-scale masterpieces in jade and pottery. A gigantic head, carved from a single piece of basalt imported from some presently unknown source, was found on the top of a hill at LaVenta.[4] Not only is the sudden appearance of such monumental statuary of previously unused materials significant, but also the features of this particular head are emphatically non-Indian. The nose is flat, the lips are thick, the eyes have a shape unlike those of the Mayan, and altogether the appearance is Negroid in character. This gigantic head also wears a helmet. No comparable head covering has been found in Meso-America before this time nor has been portrayed on any other statuary, and no metal helmet of this sort has ever been discovered subsequently or elsewhere. This particular piece of armament which finds its closest

The Mayan Synthesis • 131

The first of these colossal heads was turned up in 1939 by Hayden Stirling of the Smithsonian Institution. Since that date, heads have been turning up with considerable regularity. "The features are bold and amazingly negroid in character," Stirling said. These heads were carved from a single massive block of basalt, weighed ten tons or more, had been quarried more than seventy miles away, and then pulled or pushed to their mounting. "The workmanship is delicate and sure; the proportions are perfect," according to Stirling. All the heads are carved without necks or torsos, and are mounted facing precisely to the East. "These sculptures have no antecedent, and fit into no known aboriginal cultural picture." In view of the fact that there is no precedent in the Americas, the only alternative is that the model arrived by ship.

similarity in the helmets worn by Celtic warriors is uniquely European in its origin as much as the features of the face are African. As in the case of all similar stone carvings, erosion is inevitable and dating is difficult. However, the best assessment is that this enormous head was carved sometime between 600 B.C. and 300 B.C.

The fact that such gigantic stone sculptures had no precedent in Mayan artistic or architectural tradition is one evidence of the introduction of new cultural forces—quite aside from the non-Indian facial features. Historians and archaeologists believe that artistic and social forms grow out of the ecology of the regions in which people live. Thus, it would be natural for the Egyptians, who live in constant sight of limestone cliffs, to hew figures out of the limestone. It would also be consistent to expect them to cut the rock out of the valley walls to erect temples, some of them mammoth in proportion. In the same way, a certain logic seems to justify Gutzon Borglum in his desire to cut the faces of four great Americans in the craggy cliffs of the Black Hills of South Dakota. However, if the people of Nebraska were to go to some distant point, carve out enormous blocks of stone and convey them by whatever means to Lincoln where they would be re-erected and carved into the likeness of famous Americans, we would think that something more of an explanation would be required; and when a similar event occurred in Mexico, obviously an important cultural change had taken place. The appearance of these sculptures in Meso-America and northern South America, often appearing in plains and marshlands and built with stones that had been hauled for hundreds of miles, was a translation of skills and traditions acquired by people elsewhere, and based on memories of events that occurred in other times and other places.

The construction of pyramids in Mexico and northern South America also seems to be ecologically misplaced. The

Great Pyramid to the Sun at Teotihuacan is made of rounded stones small enough to be carried by a single person. The task of binding all these stones together must have been a considerable problem in the past and continues to be one in the present. Furthermore, in South America the Pyramid of Huaca del Sol on the Pacific desert coast at Moche was built of the incredible number of 120 million adobe bricks.[5] There is a certain logic for the Egyptians to build pyramids quarried from stone blocks weighing from four to seventy tons. However, to build a comparable structure out of sun-baked bricks of dried mud defies the simplest logic of the strength of materials, and only because this pyramid was built in an arid climate is it recognizable today as something comparable to its original structure. Had it been built in a seasonal climate, it would be unrecognizable as a muddy drumlin. At Moche, the building of a pyramid was an imported idea, possibly from the Olmec civilization of Mexico, who might have previously borrowed the idea from the Egyptians.

On the Yucatan Peninsula, buildings began to appear with corbeled walls and roofs with a single removable cap stone of the sort developed by Egyptians for astronomical purposes. Stela—intricately carved stones in upright position—performed the same function as obelisks, namely, to measure the passage of time, predict the various equinoxes and solstices, and eventually to measure the size of the earth and to map it. The fact that people did not understand the functional purpose of forms they were copying did not prevent them from reproducing such structures.[6] The Washington Monment is proportioned as a perfect Egyptian obelisk with tapered sides and a pyramid on top. Why an obelisk in Washington, D.C.? Why an obelisk as a memorial to a man who lived in a time when little or nothing was known of Egyptian culture and who probably passed his entire life with complete indifference to it? The reason is that the Washington Monument was designed just before the Civil War when Napoleon had conquered

Egypt and brought back an obelisk for Paris, and this stone monument had become the emblem of a great empire. It was necessary, therefore, that the English have an obelisk in London, and shortly thereafter the Germans had another in Berlin. At the time the Washington Monument was designed, an obelisk was the badge of an important city. None of the obelisks were erected to perform the function that was originally intended for them, namely, to show the equinoxes on the day that one side was completely illuminated and another side entirely in shadow; to provide the base point for the survey of land, or as a boundary point that separated one territory from another. In the nineteenth century, the obelisk had acquired a different function: it was the sine qua non of the capital of a great nation. So Americans built the largest obelisk in the world.

A similar cultural anomaly is found in the portrayal of lions on the sculptured reliefs of the pyramids of Chichen Itza in Yucatan. Why lions, you ask, when there are no lions in North or South America? For the Egyptians, the Abyssinians, the Persians, and Israelites, the lion had been a symbol of power, force and fear. Borrowed by the English kings, the lion became the imperial symbol because for other kings the lion had been the imperial symbol. English kings had never seen a lion, just as they had never seen a turkey before the discovery of America. To have a lion on the pyramid at Chichen Itza is a borrowed idea just as much as a statue of an anteater would be on the entrance of the U.S. Supreme Court. The introduction of such forms, shapes and symbols into the life of the Olmecs was certainly a product of trans-Atlantic migration. Although we do not have a record of each boat or raft that put off from the shore of Morocco for the new land, it is entirely probable that these new ideas were the product of colonization attempts carried out jointly by the Phoenicians and the men of Egypt.

In addition to the similarities just cited, scholars have listed sixty cultural parallels that are found in both the Egyptian and Olmec societies. In all honesty, scholars are able to complete a list of sixty cultural characteristics found nowhere except on the Nile and in the Yucatan, and still believe that no one reached America before 1942. Many of these doubting Thomases are American. It is curious that among those historical researchers who have pushed most ardently to find data, either to confirm or to deny pre-columbian exchanges, few have been American. Thor Heyerdahl is a Norwegian; Barry Fell is a New Zealander. They are two current advocates who are most strongly experimental in their effort to test the various cultural possibilities that offer an explanation for American development.

Few Mexican socio-anthropologists find any reason to believe that the Mayan development was anything other than the product of the genius of the Mayans themselves. Among those who resist the concept of cultural transfusion from Africa and from Europe prior to 1492, the strongest are in the Latin-American countries south of the Rio Grande. To suggest that the achievements of early cultures might have been borrowed from others, either European or African, is to detract from legitimate national pride. Such resistance we can all understand, and to it we can give our sympathy.

When a list of sixty unique cultural parallels is presented to them, their response is "Prove it!" Every American knows that in Washington, D.C. there is an obelisk. In classical Egypt, the obelisk has an architectural shape that was carefully designed after trial and error, to provide a variety of social and economic services, including a simple one of serving as a sun dial to indicate the time of day. The obelisk in Washington serves none of these functions. And yet the very fact that this one structure exists is evidence there has been communion and communication between this country and Egypt. Yet who is able to prove it.

If the architect for a court house in Shiawassee County,

Michigan, were to embellish that building with a frieze of kangaroos in various postures of activity and quiescence, I am confident that one fact would be evidence of contact between the designer of the building and the culture of Australia. But it is very unlikely anyone would be able to prove it. There is no architect I know, or have heard of, who is able to design a marsupial correctly and portray it in its moods and demeanors without some previous experience with marsupials. The simple fact that lions are portrayed correctly and in accurate detail on the Temple of the Sun at Chichen Itza is convincing evidence that there was at least one human contact between the people in the world where lions live and the people where lions are unknown.

As clever as the stone cutters were in ancient Yucatan, they would not have used the lion as a motif on an important public building obviously intended to convey a sense of awe and wonder in the eye of the beholder unless there were in the Olmec population a considerable number of people who had at least an inkling about lions sufficiently to regard them with awe and respectful apprehension.

Who were the Olmecs?

This is the key to the riddle.

Perhaps they were several kinds of people. Perhaps they represented several races as suggested by the cultural artifacts that exist today. Perhaps they arrived over centuries of time and married generously with the Indians who were the first residents of the central plains of Mexico.[7]

Traces of well-organized human communities are found on the gulf coast of Mexico dating back to 2000 B.C., and these dwellers were unquestionably the human predecessors of the Indians who received the various migrations of Europeans, Africans, and Near-Easterners as they found their way to the Caribbean, crossing on the Humboldt current from the Gates of Gibraltar, down the northern coast of Africa and across the Atlantic into the Caribbean Sea. Those people who occupied the gulf coast of Mexico and the Yucatan Peninsula for 1,400 years before the rise of the Olmecs are authentic subjects for the study of archaeologists because there was no written language and hence no pre-Olmec history. None of the Indian tribes from the Bering Straits in the north to the tip of Tierra del Fuego in the south developed a written language. They had iconography. They developed hand signals and smoke signals that conveyed limited but specific meaning. After 600 B.C., there were two—possibly three—exceptions to this universal lack of a written language: the Micmac Indians of southern Canada developed a system of hieroglyphics clearly based on the Egyptian system and which arrived here sometime prior to A.D. 200. The beginning of the Olmec system of hieroglyphics is less easy to date with precision, but no one places it earlier than 600 B.C.[8] The third possible exception may have been the Viracocha of the Inca Empie who might have had a form of writing kept entirely for the use of the ruling class, but which was supplemented by the quipu, used primarily for bookkeeping and mathematical recording.

The Olmec civilization was a perambulating one, and most of the central sites were occupied for no more than three hundred years. Why these centers of culture were developed to a high degree of perfection and then totally abandoned remains one of the great enigmas of human history. The city of Teotihuacan traces much of its cultural heritage directly back to the Olmecs, although that inheritance passed through the Toltec and Mixtec phases before the city of Teotihuacan, and the people known as "Teotihuacanos" who populated it had their maximum influence in the years 400 to A.D. 600. The city of Teotihuacan, which was part of the total Mayan civilization beginning with the Olmecs, came to a close during the eighth century, at which time the countryside was depopulated, the temple sites were unused and unkept, workshops and market places were abandoned, and the entire complex was allowed to be grown over with trees and vegetation.

It is one of the ironies of history that the Aztec civilization was almost totally destroyed by Cortes and his Spanish conquerors who had no knowledge the Olmec civilization ever existed. One of the first Spaniards to rediscover the Olmec civilization as Friar Ramon de Ordonez y Aquiar, who lived in the province of Chiapas joining both Yucatan and Tabasco. With the assistance of his parishoners who carried him in a sedan chair for more than sixty miles, Ordonez was among the first of the Spanish to see the overgrown ruins—splendid even though covered with decay and vegetation.[9] He was assisted in his undertaking by the bishop of Chiapas who had been faithful to the orders of the Church to destroy every document or fragment that appeared to have come from a pagan hand. The bishop had faithfully carried out his ecclesiastical instructions, and one very important document on the history of the Olmecs found its way into his hands. But before he destroyed the original document that had been entrusted to him, he carefully made a copy for himself. This was keeping to the letter of the law—although there is some question about the spirit. A prohibition of making reproductions of pagan writing would certainly have been promulgated by the cardinals of Spain and the Vatican if such an idea had ever occurred to them. But Ordonez, who devoted a substantial portion of his life to a study of this document, concluded that Mexico was discovered and settled by a leader known as "Voltan" who crossed the Atlantic at least four times, possibly to bring new settlers to the American colony. Ordonez seems clear that Votan was a Phoenician and that the first four settlements were Phoenician. His estimate of the date—which was only a surmise—was that these settlers arrived at approximately 1447 B.C.

Gradually additional manuscripts and scraps of information became available, and a description of the pioneering argonauts as "Men in Petticoats" seemed to identify them even more specifically with the Phoenicians, but suggested that the actual date of their arrival was later, perhaps closer to 1000 B.C.

Another report on the origin of the Olmec civilization that was destined to be buried until the twentieth century was written by an Italian scholar, Dr. Paul Felix Cabrera, who lived in Guatemala City, which was at that time the political capital of the Yucatan Peninsula. Cabrera continued the search for new documents and came to an independent conclusion that the Olmecs had arrived on the coast of Mexico from somewhere across the Atlantic. Cabrera suggested that the strangers were Carthaginians who interbred with native women to produce the Olmecs.[10] In view of the fact that Carthage was the Phoenicians' principal colony, there is a degree of consistency between the conclusions of Ordonez and Cabrera. Unfortunately, all of these early documents on the exploration and settlement of the Olmec country were systematically confiscated by the clergy, who had them suppressed, destroyed, or sometimes buried in secret archives. Some reappeared at the most unpredictable times as a consequence of accidental searches or unanticipated revelations.

Knowledge of the existence of the Olmec culture, coupled with an awareness that in Central America there was a vast area of several hundred square miles studded with vine-covered and decaying ruins of the most majestic and provocative architecture, generated a number of presumptive scholars, some of whom devoted most of their lives to the study of pre-Columbian America. The next, and perhaps the greatest researcher, was the Abbe Brasseur de Bourbourg who prepared himself for the task by acquiring a speaking knowledge of twelve languages and a reading knowledge of twenty more. After arriving in America, he learned two additional languages. The first was Nahuatl, the language of the Aztecs which was taught him by a descendant of a brother of Montezuma; and the second was Quiche—a written version of the Mayan language.[11]

The Christian conquistadors had, of course, done an incredibly thorough job of routing out and burning every Aztec manuscript, and the effort to decipher the scraps that remain has been largely unsuccessful. However, the survival of the Mayan manuscripts was a somewhat different matter. The Spaniards knew nothing about the existence of the lost Mayan cities and the land of the Olmec temples was wild and overgrown. Subsequent to the conquest, the cities were gradually uncovered and explored, and an impressive number of manuscripts were found. As each manuscript came to light, however, Christian priests ordered immediate destruction, not being resourceful enough to think of the invention of the bishop of Chiapas who carefully made a copy before carrying out his appointed task.

A secret collection of many Mayan books was brought together on an island of Lake Peten. However, word of this secret library somehow reached Christian ears, and almost two centuries after the conquest of Mexico such was the zeal of the Christians that General Don Martin de Vrsua was dispatched with an army to destroy the books and also to obliterate twenty-one Mayan temples in which the offending documents had been found.[12] The Spanish zealots were not as successful with the Mayan manuscripts as their predecessors had been with the writings of the Aztecs, and additional manuscripts from Olmec sources were from time to time discovered. It was on the basis of these manuscripts that Brasseur de Bourbourg hoped to do for the Mayan hieroglyphs what Champollion had done for the Egyptian.

In this effort, Brasseur was only partially successful at best. Unfortunately, he was never able to find his Rosetta Stone. If there was a key to the Olmec language—and that key would be the same text written simultaneously in two different languages, one of which was understood—that key was probably lost when all of the Aztec documents were destroyed.

The Mexican Champollion was confronted not with one system of Mayan hieroglyphs but several. Whereas in the case of the Egyptian language, the hieroglyphs had been standardized about 3000 B.C. and continued in use with no evolution either in form or grammar for more than three thousand years. Unfortunately, the same was not true of Mayan writing. Undoubtedly there were a variety of tribal languages that were translated into hieroglyphics so that the same symbol would have more than one meaning, depending on the tribe or nation using it. Or perhaps there were different systems of hieroglyphs that evolved at different times and were used by different groups as one cultural site was abandoned and civilization moved to a different one. Nevertheless, neither Brasseur nor any of his successors have ever been able to put together a neat little Mayan grammar such as Samuel Alfred Browne Mercer was able to do for the Egyptian language that made his hieroglyphs easier to understand than an elementary text in Latin.[13]

Nor was Brasseur a failure. He was able to work out an accurate understanding of the Mayan calendars, and this was an intellectual achievement of the first rank.

On the basis of Brasseur's research, he was able to systematize the utterly amazing system of Mayan mathematics which in its simplicity, clarity, and accuracy still leaves us in awe and wonder, and about which more will be said later.

Brasseur did manage to get a handle on at least a partial translation of the Olmec hieroglyphics based on his knowledge of Quiche, which is actually a language used somewhat later than the Olmec period, plus his general knowledge of linguistics. In addition to his ability to decipher the Mayan calendar system, and his remarkable achievement in mastering Mayan mathematics, one other major contribution is ascribed to him: In searching through secret archives and collections, he found Mayan records of the arrival of thirteen different expeditions that landed on the shores of the Gulf of Mexico before the time of Cortes.[14] These landings are dated and the landing places known with relative accuracy.

Two additional attempts to translate Mayan writing have

The Egyptians used both glyphs and an alphabet in their writing. After careful study, LePlongeon came to the conclusion that Mayan writing was structured along similar lines. His analysis showed that the alphabet of the Mayans was remarkably similar to the Egyptians, although the glyphs defied more than a piecemeal translation. In spite of these difficulties, LePlongeon used his knowledge of the Mayan language to make some startlingly accurate predictions. One of these involved the discovery of the Choc-Mool on the basis of inscriptions at Chichén Itzá. A group of diggers were taken to a "wildly implausible" site, and were told to dig twenty-four feet down, where they would find an important ceremonial statue. At slightly more than twenty-four feet, they made one of the great discoveries of anthropology. Contemporary scholars believe that the basin held by the Choc-Mool was a sophisticated instrument for astronomical observations.

been made and both are worthy of attention. Augustus Le Plongeon plunged himself into the unfinished task of Brasseur. By the time he was on the job, which was 1875, the knowledge of Egyptian hieroglyphics was almost complete and Le Plongeon, who was well versed in classical Egyptian, declared that one-third of the Mayan words he translated were of ancient Egyptian origin, and that the grammatical form of the two languages was similar.[15]

A recent massive attempt to translate the Mayan language has been made by scientists, historians, mathematicians, and linguists at the Novosibirsk Science Center in the U.S.S.R. After the initial construction and installation of one of the major computers of that country, a trial run was made in an attempt to use this sophisticated equipment to solve the problems that had eluded Brasseur de Bourbourg and Le Plongeon. The output of this study, which runs to a vast number of volumes, has only been seen by a limited number of Americans, and therefore final judgment must be reserved. However, it is reported that, contrary to Soviet claims, this research has added little to what is already known. In 1976, Knorosov produced a new book on Mayan writing which he believes is the long-awaited breakthrough, but scholars in other countries have not had enough time to work with his concepts to give them a fair evaluation. The fact that the Russians have committed such extensive scientific manpower and generous funding to this interest in Mayan writings makes one curious as to what they expect to find or may have already found that prompts their further search.[16]

The total collection of Mayan writings beginning with the Olmecs and continuing through the Toltecs and the Teotihuacanos is now an impressisve mass. Edward Lear, an Englishman who later became Lord Kingsboro, spent his time and fortune gathering together every Mayan document known to exist, and these were printed in a ninety-volume opus which left him in debtor's prison.[17] The final volume in this set came out about the middle of the nineteenth century, and since that time more documents have come to light, but nowhere among them has appeared a "Rosetta Stone." Meanwhile, Lear's ninety volumes of documents are waiting to be translated, and other more recently discovered writings have been distributed through the libraries of the world.

Short of a scientific breakthrough providing a new understanding of linguistics, the best key to the Olmec mind would be to continue the search for a Rosetta Stone or a Rosetta Page. The best chance of finding a pathway into the past would seem to be through the Aztec language, Nahuatl. On the other hand, since the Aztec records were totally destroyed, it seems unlikely that a bilingual document will ever come to light on the North or South American continent, and therefore the probability of finding such a semantic key would be better served by looking on the other side of the Atlantic. The Irish Celts and the Spanish Celts traveled back and forth, but since both of these Celtic regions eventually became Roman Catholic, and any Indian document was anathema, it is unlikely that any bilingual translations—even if they were made—have survived the centuries. Carthage probably maintained the largest volume of concourse with the Mayans, but this city and its people, too, were destroyed so completely that there is nowhere to search.

The Phoenicians sent original colonies to Central America, and some future exploration of the territory that is now Lebanon may produce a bilingual text. Let us hope so. Communication between the Egyptians and the Olmecs of Yucatan also existed, and the fact that the Egyptian Coptic Church along with three other Orthodox churches split off from the rest of Christianity at the Council of Chalcedon means that hostility to Indians ideas, so intense among Roman Christians, might not have existed among the Christian Copts. This toleration might have permitted documents to survive. Furthermore, new Egyptian documents are coming to light constantly, and it is possible that the Cabinet d'Egyptologie in Paris may sometime turn up a Mayan text in the process of translating papyrus writings taken from mummies exhumed from the Egyptian desert.

What we do not know about the Mayans seems mountainous in comparison to what we do. However, based on Brasseur's and Le Plongeon's lifetime of study, we do know that the Mayans developed a calendar system that was considerably more accurate and intricate than our own. They kept four calendars for the measurement of the passage of time, including the Egyptian Sothic year. One was a calendar for 26,000 years based on an accurate prediction of the declination of the earth on its axis, which involves an understanding of the North Pole and Magnetic North, and which causes gradual changes in latitude and climate. They were aware that the length of the year was 365.2420 days, and kept calendars that measured years not only in days and hours, but in seconds. [18]

They kept calendars on the movement of the planets, correctly including Pluto which was not rediscovered until early in the present century. They kept a calendar on Xiknalkan, another planet which current astronomers have yet to rediscover but which theoretically does exist, because the magnetic forces of the planetary system cannot be perfectly balanced or understood unless such an assumption is made. There were Mayan observers who with the naked eye could see the two moons of Saturn—an optical feat that today can be accomplished only by telescope—but the fact that they could accurately compute the orbit of these moons is evidence that some of the astronomers were gifted with abnormally acute vision.

The Mayans had calendars for the calculation of the solstices, the eclipses, metronic cycles, and zenith passages. They calculated eclipses into the third millennium of our era. These astronomical events, which will not occur for another

146 • *Before Columbus*

An "introducing glyph" is found at the top and spans the double row of glyphs beneath in a traditional arrangement. On the left side of the first row can be seen the figure "nine," which consists of a perpendicular bar, and a row of four perpendicular dots or circles. The bar, of course, stands for five, and each circle for one. The ninth baktun was a 33-year period of time from 613 to 580 B.C. During this period Mayan history began, and in this period occurred one of the most incredible explosions of knowledge and technology in the history of man. With three exceptions—which marked events only a few years ahead of the ninth baktun—no discoveries have been made of dates prior to this 33-year period. Curiously enough, the Mayan form of hieroglyphs appeared suddenly with no previous evolution. During this period of the ninth baktun, the Phoenician city of Tyre underwent a 13-year siege by Nebuchadnezzar. The Phoenician navy was at its greatest strength. Constance Irwin suggests: "If ever the Phoenicians attempted to establish a colony on New World soil, this might have seemed to them a propitious moment."

The Mayan Synthesis • 147

thousand years, were computed to the second, and these calculations have been confirmed by modern astronomers as accurate.

One of the best known sculptures of the Yucatan is the Choc-Mool, which was unearthed by Le Plongeon, and which is a recumbent figure bearing a basin on its stomach, sitting with its knees at approximately 45 degrees. The figure has "long ears" representative of the Viracochas of the Inca civilization, or the "long ears" of Easter Island. Visitors to Yucatan have been told, at least until recently, that the receptacle was a bowl in which the priest placed the still throbbing heart of the human sacrifice. Such a story was generally unquestioned, and satisfied the visitor that in spite of the magnificent architecture which surrounded him, the native Olmec was still a savage and that he may have been able to build a pyramid of classic interest but only for bestial purposes. The Choc-Mool, it now seems clear, was an astronomical instrument. The bowl held on the Choc-Mool's stomach was filled with water, or possibly with liquid mercury with which the Mayans were at this time acquainted, but also according to Peter Tompkins was designed to observe the "split-second observation of the transit of stars," still used today by the Naval Observatory in Washington, D.C.[19]

The fact that the Mayans were able to calculate the calendars for the planets also makes it clear that they understood the heliocentric character of the solar system and were aware that the earth and the other planets revolve around the sun. It is possible that the discovery of this important principle of celestial mechanics was borrowed from the Egyptians, or it is possible that their astronomers made the discovery independently. However, this understanding of the solar system was unknown on the European continent until almost 1,500 years later and Copernicus, one of the geniuses of all time, was to correct on his deathbed the final proofs of a manuscript which would demonstrate the same thing—that the earth circled around the sun—to the unbelieving eyes and minds of Europeans. Today, Copernicus is claimed by two or three nations which at one time or another controlled present-day Poland; but to all Europeans he is a hero of first rank who had sufficient confidence in his own observations to break into what is called the "modern day." Without distracting one iota from Copernicus' integrity, we know it would have cost him dearly if he had recovered from his final fatal sickness. Nevertheless, we who think of ourselves as Americans, even though we have come to these shores from every corner of the globe, are entitled to take pride in the fact that the home team scored such an important intellectual victory almost 1,500 years before our European cousins were unwillingly required to make the same discovery.

The achievement of Mayan astronomy would not have been possible without a system of mathematics which permitted the rapid calculation of very large numbers. Constance Irwin believes that the Phoenicians conveyed these important mathematical procedures to the Olmecs after first having acquired them from the Babylonians, but Peter Tompkins is equally certain that this information probably grew out of collaboration with the Egyptians. Certainly the Babylonians are a possible source because Babylonian cuneiform tablets now in the process of translation contain numbers as high as three trillion. C. W. Ceram points out that it was not until the nineteenth century that the concept of a million became common in Europe.[20]

Certainly the Babylonians had the concept of zero, as did the Phoenicians and as did the Mayans. The invention of zero, which has generally been attributed to the Arabs, is a subtle one. It is a figure that represents a quantity which, added to another figure, leaves the total the same as the previous figure. Likewise, zero when subtracted from a number leaves the number unchanged. As essential as this idea is to the development of mathematics—and zero is to mathematics as the wheel is to mechanics—this concept somehow eluded the Greeks and the Romans, who never passed out of the sixth grade in terms of their mathematical sophistication.

Just as language—its grammar, word structure, pronunciation, written form and syntax—is an evidence of the relationship of one culture to another, so is mathematics. Peter Tompkins is a specialist in the mathematical precision and perfection of the Egyptian pyramids. He has made comparable studies of the pyramids in Mexico, and particularly the Pyramid of the Sun and other structures built at Teotihuacan. While the Great Pyramid at Cheops in Egypt has been an object of admiration and speculation for centuries, it was not until 1925 that the precise base of the pyramid was established. That base has now been measured to the millimeter and it is precisely 500 cubits, or 750 Egyptian geographic feet. This distance, according to Tompkins, is exactly the distance traveled by the earth at the equator in half a second of time, there being precisely 86,400 seconds in one twenty-four-hour day, and 86,400,000 cubits in the circumference of the earth. The units of measurement used in the construction of Cheops were identical to those used in all other Egyptian calculations, including surveying, architecture, astronomy, and the establishment of the boundaries of the empire. These measurements were also used to predict the length of the longest and the shortest day, namely, the solstices, and as a consequence were the basis for the measurement of time which made the calculation of the various calendars possible and also the prediction of the seasons.

The city of Teotihuacan was a planned city. It was laid out in advance; and the units of measurement used in this planning were the Egyptian units of time and distance. As a basic thesis, Peter Tompkins put it this way:

> Furthermore as there appears to have been a great deal of intercourse between the Middle East and Mesoamerica, with its flow

of technicians, as well as religious and philosophical notions, it seemed reasonable to assume that any earth-consumerate unit be related to the unit used in the building of the Great Pyramid of Cheops, or at least to have been devised from some common source.

In 1972, an American engineer, Hugh Harleston, Jr., became so enraptured with the beauty and the mystery of Teotihuacan that he decided to build a mathematical model of the center of the city in order to determine what units of measurement—if any—were used; whether the same units were used in the construction of other buildings; whether these structures were related to each other in common units of distance or whether their location was a matter of pure happenstance, or perhaps some other system of measurement derived from some other source.[21]

Teotihuacan was the largest and the most impressive of the Mayan cities. It is located some thirty miles from Mexico City, and was the last of the Mayan cities to rise in eminence before the era of the Aztec. Teotihuacan at its height in the sixth century had a population of not less than 200,000, and the city covered eight square miles which meant that it was larger than Imperial Rome, and at its zenith, larger than any city in Europe. Mayans were a theocratic people and Teotihuacan was both a political and religious center. But it was more than this. It was a manufacturing complex containing over five hundred workshops for potters, weavers and sculptors, and a shopping center for the sale of merchandise not only locally manufactured, but from other countries and from other continents as well.

Laid out shortly before the birth of Christ, Teotihuacan was designed to express a unity of earthly, celestial, and possibly cosmic relationships. The ceremonial center of the city has been excavated and restored, but large portions of the city still remain slumbering under more than a thousand years of accumulated debris.

Because religion and astronomy were closely intertwined, it was a center for learning as well, and there doubtless were educational institutions to prepare the priesthood, carry on the astronomical observations, record the data, and teach participation in ceremonial functions. Unfortunately, because of he complete destruction of Aztec libraries and because we are unable to decode accurately the manuscripts we now have, we do not know or understand how the economy of Teotihuacan was organized, how distribution functioned, how the market performed, and how or whether wealth was accumulated. We do not know and perhaps never shall know their religious beliefs and practices in anything more than a superficial way, and we know little of their family and community structure, although we are able to make some shrewd guesses based on architecture, which is an enduring form of the language of a people.

Those same facets of Inca society which we can see with jewel-like clarity are blurred and often undiscernable in the Mayan, and the pages of any volume dealing with the daily life and thought of the Teotihuacano may be destined to remain eternally blank. The city fell about the end of the ninth century.

What holocaust swept the city at this time we do not know. The metropolis might have been subject to sudden military attack, but that seems unlikely in view of the fact that there was no other population centers which could produce even a portion of the manpower that was potential in Teotihuacan. Perhaps fire destroyed the city. If it did, there is strong evidence of incendiarism: the people or some group of people within the city destroyed it themselves, almost on the basis of a prearranged signal. Perhaps some great plague destroyed the people quickly, but in this case the excavators would expect to find the bodies of last survivors in the dwelling places, and this is not the case. Perhaps the land was exhausted because of long years of growing corn, but this kind of crisis requires at least a few years to consummate, so that an orderly withdrawal from the city could have occurred. Oddly enough, the walls of the dwelling places, that is the apartment houses or dormitories in which people lived, were uniformly destroyed suggesting internal revolt and sabotage. Whatever catastrophe caused the city to fall, it was sudden, apoplectic, and permanent. After a few years, some Toltecs drifted back and lived among the ruins, but overnight the greatest capital of the Americas ceased as completely as if it had suffered cardiac arrest.

Beginning with the Pyramid of the Sun at the center of Teotihuacan, Hugh Harleston, Jr. began his search for the mathematic logic of the city—assuming there was one—and if a system could be found, to determine what relation, if any, if had to other mathematic systems of other cultures or races, either contemporary with the building of the city, or known to have existed in previous eras of time.

The starting place for Harleston in his attempt to fathom the logic of the plan of Teotihuacan was the measurement of the base of the Pyramid of the Sun—the highest and most imposing structure and obviously the central point for any other calculation. One corner of the pyramid had been ripped away by a nineteenth-century excavator who as convinced that the pyramid contained great wealth comparable to Egyptian tombs. He found nothing, but he destroyed an important point in calculating the pyramid's base. Thus far, only one corner of the pyramid has been located with precision. To calculate the base, and to find the answer to other important questions, aerial photography was carried out by Rene Millon of the University of Rochester. Careful observation was made of the sun's path at its zenith. While some latitude for uncertainty must be allowed, Harleston's conclusions are that the base of the Pyramid of the Sun is 500 cubits per side, a perfect square, and this corresponds exactly with the base of the Pyramid of Cheops.

Since there is still no way of precisely measuring the base of the Sun Pyramid for which only the northeast corner has so far been found, Harleston tried to apply the logic of the Egyptian cubit to the Palenque Temple of Inscriptions for which precise baselines have been determined. Harleston used every known unit of measurement—both those currently in use and those used in all eras of previous history—but it was only the Egyptian cubit on which all measurements came out precisely, and it was the only unit for level, heights, distances, ratios, and square roots that worked with invariable accuracy. Furthermore, the application of this unit of measurement shows that it has been used in Yucatan for all pyramids, temples, and courts from approximately 300 B.C. until the invasion of Cortes.

To Harleston it was clear that Teotihuacan had been designed as a celestial clock, as an astronomical observatory, as a geodesic point for the survey of the country of which this city was capital. It was a marker to locate man in the cosmos. The design of the city was an inspired attempt to find the relationship of man to the eternal. It was an effort to give to the residents of Teotihuacan a firm place in the grand design of the universe.

Harleston is satisfied that a study of this city provides a rare insight into the working of the Mayan mind and its search for the universal. All of the subtle mathematical relationships he discovered are reported in exquisite detail by Tompkins in *The Mysteries of the Mexican Pyramids*, published in 1976.

Albert Einstein spent his final years searching for a "field theory" which would unify all the other laws of energy, matter, light, and magnetism. He was never able to formulate the equation he was searching for, but apparently his hungry mind had something in common with the architects of Teotihuacan. They too were looking for a universal law, a formula to which all other formulas are related, a key to the universe, a place to stand where all things are relevant.

10 • From Mastery To Oblivion

ABSTRACT thought is the highest level of intellectual activity, and mathematical processes are considered the purest form of abstract thought. On the basis of what we now know, we can see that the human brain was "designed" in advance to permit it to perform abstract processes of such complexity that even a generation ago these processes could not be conceived, much less explained or articulated. At the present time, we really do not know the brain's limitations. But we do know that for preeminent achievement, the processes of abstract thinking must be taught and must be learned early, just as music prodigies are started by their parents before the children reach maturity and before they can be asked if they want to study music; just as ice-skating champions today are taught by their mothers from the age of three or four; or just as Olympic swimmers are taught to swim so early that they are unable to remember the time that they did not swim. The Mayans doubtless had some system for the early identification of mathematicians, astronomers, and possibly priests and administrators in view of the fact that these functions were combined in the Mayan theocratic society.

The Mayans recognized that the capacity to do symbolic reasoning must be learned early, or it is not learned at all, or learned with great difficulty. On the basis of what we know of their society, the Mayans had a powerful drive—almost mystical in character—to discover their relationship to the universe, and within that framework, the meaning of their lives. To nurture people who were capable of leading and teaching the Mayan society in this grand enterprise they must have committed some of their best time, talent, and undistracted attention. Without this kind of devotion, they would not have produced astronomers and mathematicians who surpassed by two thousand years any parallel achievement on the European continent.

While the Mayans appear to have borrowed heavily from the Egyptians and the Phoenicians in both mathematics and astronomy, they continued to advance in both of these sciences, and nowhere is their achievement more splendid than in mathematics, which apparently was taught to and understood by the population in general. Cortes' soldiers were amazed at the speed with which merchants were able to make calculations. They were awed to discover that in the sale of products, merchandisers calculated the cost of beans not by the peck or bushel, but by the bean.[1]

The Mayan mathematical system was simple, flexible and so easy to use that it should be given a trial run as an instructional method in some contemporary American school. Their system was simple enough that a child of four could learn how to multiply, divide and obtain square roots without going through the previous task of learning the multiplication tables. Three symbols needed to be learned and they were:

$$\text{dot or } . = 1$$

$$\text{dash or } - = 5$$

$$\text{shell or } ^0 = 0$$

156 • Before Columbus

These symbols were written on a checkerboard pattern of nine or sixteen squares as follows:

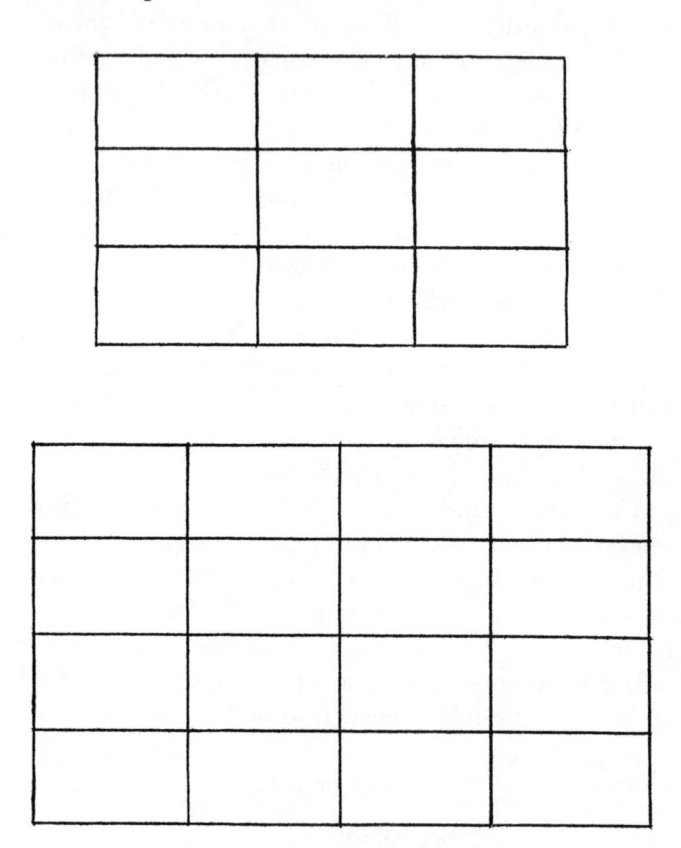

Mathematical figures were written in vertical columns reading from the top down, and reading from the left to right. Each square represented a multiplication of twenty over the square below. Thus, in order to convert Mayan figures into their Arabic equivalent, notations in the lowest square are multiplied by 1; in the second lowest square by 20; in the next higher by 400, and in the case of a 16-square checkerboard, the top square is multiplied by 8,000. The total of all of these squares is added to convert the figure into the numerical equivalent. Thus the Mayan equivalent of 16,720 would be written as follows:

and would appear in each square of the first row of the checkerboard. The Mayan notation could be converted into our numerical system through the following process:

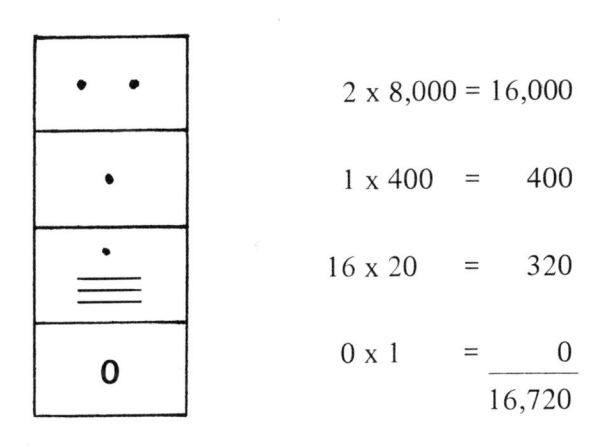

$$2 \times 8{,}000 = 16{,}000$$

$$1 \times 400 = 400$$

$$16 \times 20 = 320$$

$$0 \times 1 = 0$$

$$16{,}720$$

Now to 16,720 let us add two numbers, the first of which is 41,276, which, according to the Mayan system, would be noted as follows:

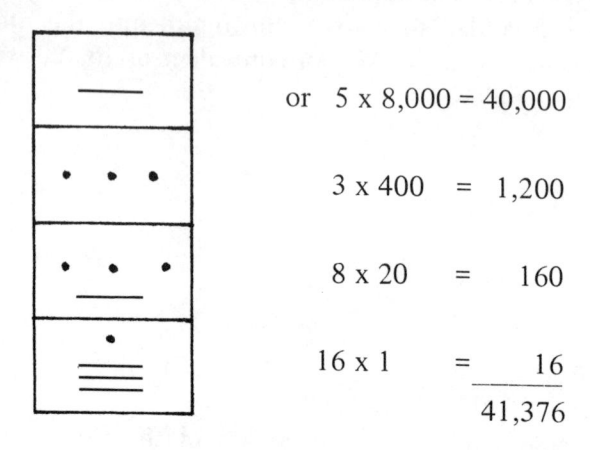

or 5 x 8,000 = 40,000

3 x 400 = 1,200

8 x 20   =   160

16 x 1   =   16
              _____
              41,376

—and an additional number 3,499, which the Mayans would note thus:

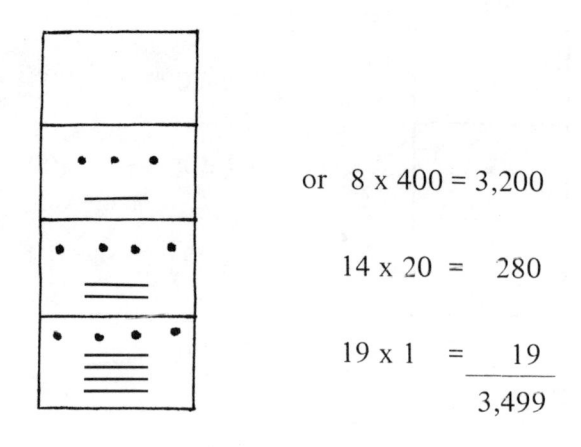

or 8 x 400 = 3,200

14 x 20 =   280

19 x 1   =   19
             _____
             3,499

Addition requires that the bottom line be added, and that each 20 be transferred to the line above as a single dot, or that five 20s be transferred up as a dash. The sum total of the three figures thus would be:

・ ・	・ ・ ・		・ ・ ／
・	・　　・	・ ・ ・ ／	・ ・ ・ ／／
・ ／／／	・　　・	・ ・ ・ ・ ／／	・ ・ ・ ・ ／／
0	・ ／／	・ ・ ・ ・ ／／	／／

or 56,000

5,200

380

15

61,595

The Mayan system has striking similarities to Boolian algebra which provided the mathematical tools Norbert Weiner needed to develop the computer. In Boolian algebra, there are two symbols, 1 and 0, and with these two symbols, Boole the Irishman demonstrated that it was possible to carry out any calculation, even the most complicated, through the simple process of addition and subtraction. The Mayan system of mathematics contains three symbols: 0, 1, and 5. However, the symbol for 5 is a convenience and could be eliminated, and on some documents was eliminated, using five dots instead, But the fact that the Mayans were able to achieve high-speed calculations involving large numbers may be the result of the fact that their system was so close in its simplicity to computer technology used in the last half of the twentieth century.

The brilliance of Mayan mathematics and astronomy is recognized by scholars as surpassing achievement in Europe or America prior to this century. However, it is claimed by scholars that the fact Indians of Yucatan were in possession of such magnificent astronomical and mathematical systems does not mean this achievement was the product of a crosspollination of ideas brought by voyagers from Europe or the Mediterranean. This intellectual accomplishment, it is asserted, might have been a natural evolution of the thoughts of native Indians on their own and by themselves.

If this was a natural outgrowth of indigenous Indian culture, then the achievements of the early Mayans must be considered one of the great cultural explosions of all times—if not the greatest in the history of man. How all of this intellectual capital could be amassed in so short a time, and starting as close to ground zero—as we know from all the other Indian societies that surrounded them—must be one of the truly breathtaking achievements of the human mind, and requires us to ask what were the forces that generated this evolutionary surge? Why did it happen in a relatively limited geographical area on the Carribbean and not among the other and older Indian tribes that carried the same genetic stock? How was it possible for the Mayans to jump so many intermediate levels of development and proceed directly—perhaps in the course of a hundred years—to a level of growth that Mediterranean and European people were able to achieve only after groping for hundreds and sometimes for thousands of years?

Not only was there a sudden quantum leap in the theoretical disciplines of mathematics and astronomy, but a new engineering also appeared based on these theoretical models, and a new architecture, economy, social structures, agriculture, and religion. How these new forms emerged without going through the trial-and-error process, as the Egyptians did in perfecting their concept of the pyramid, leaves us breathlesss with wonder. How the Mayans organized their agencies of discovery, their channels of transmission, their hierarchy of learning to achieve such a profound breakthrough in knowledge requires some explanation, or at least examination on the part of those who believe that Indians with the theoretical resources of our early Iroquois or Chippewa could within a few hundred years and without outside stimulation begin to calculate the orbit of the moons that circle Saturn.

Perhaps the incredible achievement of the Mayans was wholly a product of the internal energies and psychic tensions of the native Indians that inhabited Meso-America. This possibility ought to be studied by our most gifted scholars because, if it actually happened, the process of this achievement may provide a key to some of the most baffling problems that face mankind today. At the same time we must continue our examination of an alternative explanation: namely, that the unbelievably rapid evolution of intellectual and practical systems in Mayaland was the product of the introduction of foreign cultural elements that fused with the native Indian to produce a hybrid mix of unusual intensity and vividness.

If this later theory is correct, then there are certain areas in which we should look for evidence to give it substantiation. One is mathematics. Do the mathematical systems of the Mayans show the inclusion of elements of the mathematics of other societies from other parts of the globe? We know that in our own society such borrowing occurred, and that up until the first World War geometry was taught in the schools using the original propositions of Euclid, a direct borrowing from the Egyptians; algebra was a gift of Moors who called it *al-jabr*; differential equations come from France; and calculus was the product of the mind of an Englishman, Isaac Newton.

Secondly, we should look at language which was its own internal logic and appears to change according to given laws. Here we are terribly handicapped, because the ability to read the Mayan language has been lost, and no more than a third of the hieroglyphs are comprehensible to us today. With all the handicaps we have, is there evidence of language elements that appear to unite the Mayan with languages used by other peoples at other times?

Biological and anatomical information may give some clue as to the origin of culture. In the same sense, artistic and architectural forms may be fingerprints that allow us to trace the origin of institutions and inventions. Is there evidence in any of these areas that provide clues as to the source of Mayan culture?

As much as some historians and anthropologists would like to deny that any white man ever found his way to American shores before 1492, we are confronted with the fact that drawings and sculptures of men with beards have been appearing in Mayan culture since before the time of Christ. Apparently all purebred Indians in both North and South America have blood type O. This blood type is carried on the X chromosome as a recessive trait, which also carries beardlessness as a recessive trait.[2] No Indian of pure blood, now or in the past has had a beard, and the presence of facial hair is an evidence of genetic material other than Indian. The portrait of bearded men has occurred in various chapters of Olmec history, and one sculpture is so outstandingly Semitic in character that it might very well have been taken from a frieze on the walls of the City of Persepolis. Perhaps the sculptor himself came from Persepolis, or perhaps the stone was carved in Persepolis and brought to Mexico. There are a number of alternatives, but one thing is certain: this portrayal is not an American Indian.

Backed by the personal fortune of the American tobacconist Pierre C. Lorillard, a brilliant young Frenchman and girl-watcher in his early twenties, Claude Charnay, set out on a twenty-year mission to find the city of Tula which had been a glittering capital under the regime of the Toltecs. In 1880 Charnay did uncover a great city with a princely courtyard full of remarkable stonework and sculpture. However, his contemporaries tended to dismiss his claims as they also dismissed the claims of Heinrich Schliemann, a contemporary who was claiming that he had discovered the ancient cities of Troy and

All native American Indians from the Aleutian Islands in the north to Tierra del Fuego at the tip of South America have blood-type "O," and one characteristic of this genetic makeup is beardlessness. Thus, the appearance of a bearded person anytime before Columbus would be evidence of the introduction of genetic material not present in native Indian stock. Four samples of sculpture or pottery with bearded representations are shown above. Note also that the facial characteristics do not seem to conform to the features of beardless Mezo-Americans portrayed elsewhere in sculpture, painting, and codices. However, the most influential and beloved visitor to come to these shores is not portrayed above. He was the yellow-haired and bearded Quetzalcoatl.

Mycenae. Professional archaeologists vigorously denied that Charnay had found Tula, but gradually their conviction began to diminish.

In 1940, the Mexican archaeologist, Dr. Jimenez Moreno, agreed that Charnay was right. He had found Tula, and it had been built about A.D. 900, well after the destruction and abandonment of Teotihuacan. Tula in turn was destroyed by the Aztecs, who carried off much of its stone and sculpture to incorporate in their own capital which was later captured by Cortes. However, in the center of one wall of Tula in the main courtyard and garden, Charnay found a bas-relief which depicted two bearded men. This finding puzzled him because they were both clearly non-Indian and, although much has been learned about Tula since its discovery by Charnay, no light has been shed on the identity of the two bearded strangers.[3]

More recent research gives evidence of the arrival of foreigners almost 1,400 years before the construction of Tula, however. Near LaVenta, which was also a Toltec city, a 14-foot stele, or a stone marker, was raised to commemorate an event of importance or to designate a boundary line. On this column appears the likeness of an impressisve man whom anthropologists fondly call "Uncle Sam," because of his similarity to World War I poster of stern, bearded uncle pointing his "I Need You" finger. This sculpture has a spade beard and a high nose; however, his clothes are un-American. He is wearing a double knee-length gown, appears to have ribbons in his hair, and is sporting sandals with turned-up toes. This garb raises the question of what time in human history men could have worn turned-up shoes, spade beards, and the peculiar form of dress portrayed on this stele. A careful study of the sartorial traditions of mankind was made by Constance Irwin who found these same stylistic elements to be characteristic of the Phoenicians about 500 B.C.[4]

Of all of the mysteries of the bearded men who have influenced Mexico, none is greater than that of Quetzalcoatl. According to the Aztec legend he appeared during the previous Toltec period; and indeed the Toltec reports do agreed with all the principal points of Aztec legend. With the passage of time Quetzalcoatl became a god himself and was worshipped by the Toltecs and later by the Aztecs, and great temples and pyramids were raised in his honor.

Quetzalcoatl arrived on the Gulf Coast—apparently in the company of other men—and traveled inward to the centers of Toltec civilization to become a great inspiration to the people who in turn came to love him. He developed superb artisans, devout worshippers, skillful tradesmen, stone masons, carpenters, bricklayers, and workers in feathers and ceramics. He was a tall and virtuous man who sang and danced, and instructed the priests charging them with the keeping of an accurate account of the days and the years and the movement of the stars and planets. He believed in one god, was opposed to human sacrifice, upheld fidelity, taught virtue, was honest, believed in kindness, and emphasized morality in dealing with others.

Although he was loved, he was not without his distractors. Among the most influential were the priests who practiced human sacrifice and were now experiencing the wave of resentment against their rites. The priesthood devised their own method of undermining Quetzalcoatl. One was to design a new god, or perhaps to elevate an obscure god to the position of a strong challenger of Quetzalcoatl, and that was the God Tezcatlipoca, or "Smoking Mirror," who upheld as virtues the vices that Quetzalcoatl decried. Furthermore, the antagonistic priests attacked Quetzalcoatl charging him with a sexual offense, the exact character of which is unknown. Whether the offense was real or imagined, we shall never know; however, both written and legendary accounts of the incident are strictly on Quetzalcoatl's side and insist that he was framed by his rivals.

Quetzalcoatl felt that his reputation had been sufficiently damaged by these insinuations that he could no longer be effective and therefore returned to the Gulf with a group of friends, and set out to sea promising to return in the year One Reed. And so this white-skinned, yellow-haired, bearded man left Mexico from the same coast on which he arrived. But during his stay, however brief it was, he had a profound effect on Mayan civilization. The year One Reed on which he promised to return was, according to our calendar, A.D. 1519, and it was in that year that the ships of Cortes appeared in the Gulf of Mexico. Up to the time of Cortes, Quetzalcoatl probably was the most notable and effective individual in Meso-American history.

But who was Quetzalcoatl—this good man with yellow hair, blue eyes, and a beard? Was he a sailor who managed to make it ashore on some wreckage of the ship after a storm? Was he a refugee? His yellow hair and beard do not suggest the Mediterranean. Is it possible hat he was a Christian missionary who crossed the sea in an open boat very much on the same ocean curents that carried Thor Heyerdahl? He taught crafts; he brought a new religion; he was opposed to human sacrifice; he believed in one god; he created virtuous men.

He was known as the "Feathered Serpent," an odd name for a missionary, but if he were a Christian misssionary he might have taught his people to "Be as wise as serpents and as harmless as doves." As the Toltecs thought about this combination of being harmless as a dove and wise as a serpent, they might naturally conceive of a "Feathered Serpent" combining the two virtues, symbolic of the man and his teachings.

Speculation about Quetzalcoatl and who he was began immediately after the Spanish conquest. The Franciscan friars who began their travels to Mexico during the year 1520 were fascsinated by the origin of this man, and as they studied the rocks and temples they thought they could find crosses and

Jewish epigraphy. They came to the conclusion that Quetzalcoatl was St. Thomas—the same "doubting Thomas" who put his fingers in Jesus side. It is recorded that as a disciple, Thomas prayed to God to be sent anywhere to preach the gospel except to the Indians—and he was sent to the Indians. According to Jacques Lafaye, who has studied Quetzalcoatl and his contributions to the Mexican sense of identity, it is reported that for two centuries the idea that Quetzalcoatl and Thomas were one strongly infused Mexican thought: and it also seemed to redeem Christ's promise that "there is no man without a witness," which otherwise appeared to have neglected North and South America for a millennium and a half.[5]

But whoever Quetzalcoatl was, his goodness shines forth. He never made any claims to deity, and there is no evidence that he ever thought of himself as anything other than a man—a man who came to teach other men to be good. But goodness can be offensive. It can generate resistance, and this was evident in the persecution that was heaped on him during his life. Furthermore, even those who believe in goodness find goodness hard to practice, and many who start to practice eventually cease to try. From this sense of failure some relief is needed, for the teaching of goodness introduces a strain in human living—both for those who believe and those who belittle. For the Mayans, this strain was relieved by gradually moving Quetzalcoatl out of the realm of manhood and into the realm of godhood where he had infinite wisdom, infinite strength, infinite power, and unending life. Why shouldn't a being of such superlative qualities be good? And who is there to say that any lesser persons should not expect to fail or even fail to try? So with the passage of time, Quetzaldoatl's magnificence increased and his influence diminished.

11 • Montezuma's Testimony

THIS book is concerned with the exploration and colonization of the Americas before Columbus, and also with the reasons that lay behind the expeditions, and consequences that grew out of them. Cortes set foot on the American mainland thirty years after Columbus entered the Caribbean Sea. However, certain aspects of the conquest of Mexico throw light on the pre-Columbian past, and it is for this reason that a return to Cortes is required here. Of particular significance was the fact that he arrived on the shores of Mexico in the year One Reed which, according to the Aztec calendar, was the holy year in which the beloved man-god Quetzalcoatl had prophesied that he would return. The fact that this prediction appeared to be coming true paralyzed Montezuma's capacity to make a decision as to the character and intention of the white-skinned, bearded visitor, and of course when Cortes discovered that the ancient legend of Quetzalcoatl could be applied to himself, he exploited this belief to the maximum demoralization of Aztec defenses. Cortes was one of the brilliant and evil men produced by European Christianity, and if an effort is ever made to weigh whether Christianity has contributed more to man's suffering than it has to his succor, then certainly the unique impact of Cortes has to be placed in the balance. Cortes, in his own words, came here "to convert the Indians to Christianity and to teach them to accept God's will." There is no reason to doubt his sincerity.

After some initial misunderstandings about his authority and right to command, he received the enthusiastic and unfaltering support of Charles V, the Holy Roman Emperor who applauded and rewarded his efforts in Mexico. After the initial conquest of the country was over, those Indians who survived were brought into the Church by any and every conceivable means, but the cost of bringing Christianity to the native population was a heavy one for the Mexican people. The city of Tenocholitan when Cortes arrived had a population of 300,000. When it fell, it had been reduced to 60,000.[1] With Mexico City safely under his control, Cortes pursued the oc-

Those who believe that America was never visited before the time of Columbus point to the fact that the Western Hemisphere was ignorant of the wheel, and certainly any visitors from the Mediterranean would bring knowledge of such an invention. However, in 1940 Hayden Stirling of the Smithsonian turned up wheels in his excavation of Tres Zapotes in Vera Cruz. Wheeled toys and tiny wheeled chariots had been turned up earlier by Desire Charnay. The idea of the wheel had been known in Mezo-America, possibly as early as 500 B.C., and yet the knowledge of this device had never been applied to warfare to produce the wheeled chariot. Why? Possibly because there were no draft animals in America—no horses, donkeys, burrows, oxen, or cows. As one observer remarked, "The horse comes before the cart."

cupation of the rest of the territory. As a result of combined military action, famine and plague, the population of Mexico was reduced to approximately one-twentieth of what it was when the Spaniards first stepped on Mexican soil.

The process of eliminating the Indian people was actually begun before the fall of Montezuma himself. On the march from the coast to Tenochtitlan Cortes passed by the city of Cholula and was greeted by pretty girls singing, dancing, and playing instruments. These maidens garlanded with flowers formed an honor guard to escort Cortes and his army of 120 men to the center of the city where they were greeted by nobles, chieftans, and all of the residents of the city. A banquet had been prepared and the Catholic priests reported that the people were happy and eager to hear what the white god and his comrades had to say. On signal, Cortes had his army seal off all the entrances to the square. The people were unarmed. The little band of 120 Christian soldiers fell upon them and killed 6,000 Cholulans that heroic day.

When the country had been pacified, Cortes proceeded with enormous vigor to Christianize the portion of the population that remained.[2] As a persuader, thirty-four men and women were hanged and beheaded and their heads mounted on spikes in Mexico City. Indians who refused to join the Catholic Church were horsewhipped and humiliated, and for failing to attend mass they could receive a hundred lashes. The treatment of the Indians by the Christians had two consequences: it brought many conversions to the Church of Jesus Christ where they were taught to accept God's will, to be gentle and loving, to return good for evil, and to do good unto those who hated and spitefully used them. A second consequence of this effort at conversion was an outbreak of mass suicides from those Indians who could not bring themselves to be subservient to the white man they hated, and who refused to give the Christian the satisfaction of one more atrocity.

Cortes justified his heavy hand on the basis that Indians

had practiced human sacrifice, and that he was in fact saving human lives. In this set of circumstances, it was necessary "to be cruel in order to be kind." Apologists for the conquest have made much of the fact that human sacrifice was brought to an end, and it was their estimate that perhaps as many as 50,000 people had died annually in this cruel religious practice. However, it should be remembered that the history of the invasion was written by the victors and not by the vanquished. Later scholars who felt less obligation to maintain the sanctity of the Spanish, or the Christian mission, have seriously questioned the accuracy of the number of human sacrifices that occurred prior to the invasion. Cortes himself stated that in all his years in Mexico he had never seen a human sacrifice, and that the only ones that he had ever witnessed were merely executions of prisoners of war during siege.[3]

On his return to Spain, Cortes reported in person to the Holy Roman Emperor Charles V. He gave an explanation of Mexican history as presented to him by Montezuma, lord of the Aztecs, and a full report of Cortes' account is retained in the Spanish archives. Montezuma had said that his ancestors had come from a land far away where there were high mountains and a garden inhabited by gods. "We [the Mexican rulers], were brought here by a lord whose vassals all of our predecessors were, and who returned here to his native land. He afterward came here again, after a long time, during which many of his followers who had remained had married native women of this land, raised large families, and founded towns in which they dwelled. He wished to take them away from here with him, but they did not want to go nor would they receive or adopt him as their ruler, and so he departed. But we have always thought that his descendants would surely come to subjugate this country and claim us as their vassals."[4]

And thus we have the official history of the Aztec people. In capsule form Montezuma repeats what other historians have said, what that portion of the Mayan language we can understand seems to say, and what is reaffirmed in legend and saga. Visitors from overseas came to Mexico. They mingled and married with the people they found on American shores. Other ships came and departed. But settlers stayed—and all of this before the days of Columbus. And to this day, not one codex or historic text; not one stone upon another; not one stele or sculpture; not one artifact or fragment has yet appeared to dispute the accuracy of the testimony of Montezuma.

12 • Vikings: Moody Adventure

IN the Persian fairy story of "The Three Princes of Serendip," the heroes had the fortunate trait of finding valuable or agreeable things they were not looking for. The story of Bjorni Herjolfsson, a young Viking lad, is not a fairy story, but in terms of discovering something he wasn't looking for he was a true "Prince of Serendip." Bjorni set off in his boat from Norway in the year 986 to visit his parents in Iceland. He was a merchant; his boat carried a crew of approximately thirty-five; and he was acquainted with the ways of the sea. Surprising news was awaiting him when he arrived in Iceland: His parents had migrated with a colony of approximately three hundred toward two possible sites in Greenland that had been subject to thorough reconnaissance by Eric the Red.[1] Bjorni did not know to which of the two colonies his parents were destined, but Eric the Red had spent three years traveling around much of the island, and came back with a description of the seacoast of Greenland, including detailed descriptions of the two ports to which they bound, and the kind of land that was available for farming, husbandry, and community development.

Bjorni had already traveled a thousand miles to see his parents and he was in no mood to turn back. After receiving sailing instructions and detailed descriptions of the land to which his parents were borne, he refused to discharge his cargo, and when his men asked him what he intended to do he replied that he would keep his custom of passing the winter with his parents. "And I will," said he, "take my ship on to Greenland, if you will accompany me." They agreed, and a full account of the journey is contained in the *FlateyJarbok, Volume 1.* Bjorni, like all good skippers, kept a daily log of his voyages and much of the material in the *GraenLendinga Saga of the FlatelyJarbok* is derived from this source.

After the first few days at sea, Bjorni and his crew were lost in fog. Borne by winds that carried them for days in directions they did not know, they were uncertain where they were or how far they had gone. When the skies cleared, Bjorni sailed due west for one day and came to a shore which he described as "being without mountains, well timbered, and there were small knolls upon it."[2] This, he was sure, was not the Greenland described to him before his departure from Iceland, and he changed his course and began to sail north and back again to the east. He did not go ashore because it was already late in the fall, and he was unsure how far he was from his father's home. He knew that every day would count in getting to the fjords before they were frozen.

Two more days at sea brought his ship to another land. He had been told there were glaciers on Greenland but there were none on the shore that lay to port side, so he continued his travels on to the northeast. A third time he saw land, and he circumnavigated this island and again declined to stop. He persevered in his voyage until finally he reached the shores of Greenland, which he recognized from the accurate description, and fortunately was able to land close to his father's home. According to the saga, he settled there and gave up voyaging from that time forward.

Fortunately the dates on that voyage are clear and there is no question that the first Viking of record to discover America—although he never stepped ashore—was Bjorni Herjolfsson. According to the records, Bjorni, after three days' sail from Iceland, was driven far south for many days mostly in fog and by a northerly wind. After he turned back, he sailed nine days at an estimated standard speed of 150 nautical miles per day. In addition, he sailed around an island which perhaps was Newfoundland, and this would have added an additional 250 miles to the journey. Altogether, after first sighting the shores of America, Bjorni sailed northeast a total of 1,600 miles before he touched the shores of Greenland, and fortunately he was able to land close to the homestead of his parents. There is dispute about the three points along the Atlantic seaboard that Bjorni viewed from his boat; but in terms of the distance traveled before he reached his parents, and in terms of his description of the land itself, he appears to have sighted America first in the vicinity of Cape Cod. The second point may have been Nova Scotia, which quite accurately matches Bjorni's description, and the third was probably Newfoundland. Fortunately, the date for this voyage is the same as the date of the founding of the Greenland colony by Eric the Red. Events that were to follow shortly thereafter confirm the accuracy of these dates and events, and historians on both sides of the Atlantic are willing to accept the dates as dependable.

This does not mean that Bjorni Herjolfsson was the first Viking, or even the first European, to discover New England, as the earliest Norwegian settlers in America themselves were to make clear somewhat later. However, he was the first man "on record" in Norwegian history to discover America. It is a tribute to his filial devotion that he pushed on with his journey rather than take time to step ashore, and as he traveled approximately 1,300 miles along the North Atlantic seacoast he stored up many facts and details to report to family and friends when he touched shore in Greenland.

Bjorni's travels were a great venture, and he came from a tradition of great adventurers. Beginning in the ninth century, the Danes, the Norwegians and the Swedes put out in their boats to explore the known world and the unknown. They traveled south on pirate raids to ravage the seacoast towns of

Scotland, Ireland and England. Later they came to settle in England, and still later England was united to the Scandinavian peninsula and to Denmark under the rule of the Danish kings. Their travels took them farther: Norsemen journeyed up the Seine and twice captured Paris. They continued around the Spanish peninsula and entered the Mediterranean. They sailed from Byzantium through the Black Sea to Kiev; crossed land to the Volga River; sailed down the length of the Volga River to the mouth; crossed the Caspian Sea; and continued over land to the city of Baghdad. In comparison to this itinerary, the sighting of unknown land somewhere east of Greenland was not a portentous event.

Bjorni left Norway for Iceland, and Norsemen had known of the existence of Iceland for years, or perhaps for centuries. Not until 870 was a permanent Norse settlement attempted, and was completed in the span of sixty years, during which time the population grew to 10,000. However, digging around ancient sites showed that the Romans very likely knew of Iceland in the third century because four Roman copper coins from the period A.D. 270-305 have been found among ruins in the south portion. This, according to Magnus Magnusson, is the time when the Roman navy in Britain was at its height under the command of Carausius, and he believes it is not an unreasonable assumption that the coins were left by sailors or marines on some long-range patrol.[3]

The first permanent Norse settler in Iceland was Ingolf Arnarson who sailed west a thousand miles from Norway's shores. The spirit of adventure typical of Norse sailors is exemplified by Arnarson, who proceeded without any preknowledge of what his mission involved to sail around the entire island of Iceland into the Arctic Circle and back again to his point of departure—an added distance of more than a thousand miles.

The original settlers of Iceland were pagans, or Celts, and they exhibited the rash, independent, free-wheeling bravery, mixed with stout individualism which is inseparable and indigenous to Celtic culture. Stirring among these early Celts may very well have been a tincture of Celtic Christianity with an intensive evangelical urge to convert other Norse to Christianity, a drive which was not to reach its peak until the next century. Even after the year 1000 some of the stoutest of the Norse who founded Greenland had not been converted, including Eric the Red, who gathered together the first twenty-five ships that carried three hundred original colonists from Iceland to the two Greenland settlements.

The adventure of the Vikings as explorers has unfortunately overshadowed other important contributions made by the Norse settlers. During the rule of England by the Danish throne, a new set of laws were introduced which considerably democratized the English practice then in existence. It was from the Danes that the British acquired the idea of trial by jury which has since been a cherished part of the British constitutional system. It was the Icelanders who developed the first parliament in Europe, the "Althing," which has continued as the legitimate government of Iceland to the present day. The Norwegians in Normandy achieved a reputation for efficiency and fair-dealing. They were exemplary in terms of fair practices in other provinces and in France. From the Danes, Europe acquired not only a greater knowledge of the world, and a revivification of the spirit of adventure, but also a sense of fair-dealing that has continued to be a living tradition among those people who were influenced by the Norse way of life.

Bjorni Herjolfsson was part of this tradition when he set off from Iceland to Greenland—a country he had never seen—and by accident stumbled on the shores of America. After he returned to his family, he had stories to tell which set spinning the heads of those gathered around him and lighted up the Norse imagination and lust for adventure.

The fires of the new discovery burned hotter nowhere than

in the brain of Leif Erikson who, in addition to the Norse hunger to see new lands, thought he saw a very practical and indeed profitable dimension to Bjorni's story. In the course of this trip, Bjorni had traveled by shores that were well forested. Wood was the one important commodity Greenlanders lacked. They had been required to build their homes out of sod and stone. Willow bushes and small scrub trees could occasionally be found on the island, but were useful for little more than kindling. But wood—trees, real logs—this would make possible great construction, the kind of community growth that Greenland desperately needed. So Leif bought Bjorni's boat, perhaps on the assumption that boats know their own way back—and in the year 1001 set out with a crew of thirty-five. He followed Bjorni's sailing instructions, but landed in Labrador, which he regarded as utterly worthless. The shores of Labrador are among the most hostile in the world and Leif had no desire to linger here, and so he continued on and there is evidence that he next stopped on the northern tip of Newfoundland. This was a pleasant climate, a land of forests, and much to Leif's liking. The streams teemed with fish and there was evidence of game, so he called it Markland and until the sixteenth century it was so known in the Scandinavian countries.

To Dr. Helge Ingstadt goes the distinction of first locating the foundations of a Norse colony in Ameica founded at L'Anse aux Meadows, Newfoundland, but possibly not by Leif Erikson.[4] She and her colleagues have excavated the site of two great houses that correspond very nicely to the abandoned Norse dwellings that have been uncovered in Greenland. One of the two houses is 70 x 55 feet and attached to it are rooms for family living. The very special shape of the building, which has a central area for fire with an opening in the ceiling for smoke to escape, has draped family rooms around the edge. It is typical of houses built at the same time in Iceland, and also in the Norwegian homeland. Furthermore, this type of building the Vikings also carried to England, and portions of such structures still remain on the Isle of Man and other centers of Norse influence. Dr. Ingstadt estimates that between seventy-five and ninety people were able to live here at one time. The explorations which have been carried on over a span of years up until 1968 has also resulted in the discovery of a number of cultural artifacts that identify it unmistakably as an eleventh-century Norse settlement.[5]

The climate of Newfoundland was not sufficiently benign, however, to produce the one crop that was ultimately to give the new country its name—wild grapes! Furthermore, Leif knew that according to Bjorni's travel distance he had at least one more port to make, and so with his colony he set off again toward the general area of Cape Cod.

A description of the landing of the Europeans on American soil is given in the *GraenLendinga Saga* as recorded in the *FlateyJarbok* as follows:

> They went ashore and looked about them. The weather was fine. There was dew on the grass and the first thing they did was to get some of it on their hands and put it to their lips. And to them it seemed the sweetest thing they ever tasted.
> There was no lack of salmon in the river or the lake, bigger salmon than they had every seen. The country seemed to them so kind that no winter fodder would be needed for livestock. There was never any frost all winter and the grass hardly withered at all.[6]

Leif divided his group into two exploring parties of equal size. On one day the first group would go out and the other would remain at home to build the accommodations for winter. The following day, the arrangement would be reversed, and the explorers of the previous day would remain at home while the second group went out to seek and to find. On one of these visits, a German by the name of Tykir made an intoxicating discovery: he found grapevines growing wild. He came back and explained that there was no question about his discovery. He had lived in Germany where grapes were grown. He knew the character of the vines, and he knew how wine was made. He gave full assurance that these were authentic grapes and the quality of the vintage would be delightful and thirst-quenching. As a consequence of this discovery, Leif Erikson called the new country, "Vinland" or "Wineland." To the other virtues of the Norsemen, the art of salesmanship probably should be added. It was Leif's father who had given Greenland its name, because it sounded so much more attractive than "Iceland." With a real flair for public relations, Leif knew that to call the new country "Wineland" was to pick a winner.

Leif stayed in Vinland that winter and probably for most of the following summer when he loaded his boat with all the cargo it could carry, including precious materials such as wine and lumber. On his way back to Greenland, he had the good fortune to rescue the crew of a wrecked ship, and among the people he saved was a woman called Gudrid, who played an important role in later Vinland adventures. The discovery of a wrecked vessel proceeding to America on an expedition which is not recorded in other sagas or literature, suggests more exploration of the American coastline than the voyages recorded in the few written sagas which have survived to the present time. The good luck of finding the foundering ship and rescuing the survivors added to Leif's fame. He was thereafter nicknamed "Leif the Lucky."

Another and more substantial effort at colonization occurred in the year 1010 when a man of considerable wealth, an Icelandic merchant named Thorfinn Karlsefni came to Greenland, spent the winter in the superlative household of Blattahlid, and there met Gudrid, who had been saved by Leif the Lucky when her ship was wrecked on the way to America.[7] She is described as a woman of great beauty and social grace, and Karlsefni quickly fell in love with the attractive and eligible Gudrid, who married him and accompanied him to America where they used Leif Erikson's buildings for permanent settlement. In the expedition were 160 men and five women, one of them Gudrid. During the first year Gudrid bore Thorfinn a son named Snorri, who became the first child in the new colony.[8] Therefore be it remembered that not Virginia Dare but Snorri Thorfinnson was the first white child born in America.

This colonization lasted only three years. Thorfinn may have been an excellent merchant, but he lacked a certain tact and a sense of human nature: one hundred and sixty men and five women is a certain formula for disruption.[9] After the first year, the men began to quarrel about the women, and the men also began to cheat the Indians with whom they traded. The colony was therefore divided within and persecuted from without. They decided their only alternative was to return to a more normal standard of living, which meant to go back to Greenland and make the better part of an unhappy bargain.

From Greenland, Thorfinn and Gudrid returned to his substantial estate in Iceland where he predeceased her, leaving her with considerable wealth. She managed to overcome her grief, to live an impressive and exciting life, and to travel to Rome and elsewhere in Europe.

The *GraenLendinga Saga* records three additional expeditions to the Vinland, but certainly a larger number than this occurred. The skill and daring of the Norse mariners had continued to improve and it became the practice to sail directly from Greenland to the coast of Norway, a distance of approximately 1,300 miles, which under normal circumstances would require about ten days on the open seas. In addition, Greenlanders found that one of their most valuable resources was walrus tusks which were of such excellent quality that for a period of time their trade threatened to eliminate elephant tusks from the European market. It was necessary for the Greenlanders to travel beyond the Arctic Circle in order to harvest this profitable crop, and it was their practice to build small refuges where they temporarily housed themselves for these tusk-gathering expeditions. Many of these habitations have been found, and in 1880 the Danish navy discovered one as far north as a thousand miles from the closest Greenland settlement.

For five hundred years before Columbus, the Viking set off across the North Atlantic to Iceland for a voyage of 1,300 miles, a distance almost the equivalent of the shortest route from Africa to Brazil. He then continued to travel a thousand hazardous miles into the frozen North to search for walrus tusks. Would such a daring adventurer risk a trip of two hundred miles or more to explore the more appealing seaboard of America? The answer is that he would, and that he did.[10]

Two interesting curios were found in the home of Gudrid and Thorfinn, apparently souvenirs of their three years in Vinland. One is a piece of anthracite coal and the other is a quartzite arrowhead. The anthracite coal, of course, could have been an import from Europe. However, it is a lump that can also be identified with the anthracite found in Rhode Island. There is no question about the quartzite arrowhead, which is of Indian origin.

The *GraenLendinga Saga* contains a number of other interesting historical anecdotes, but one of the most important was the fact that the Norse who established the original colony discovered that they were not the first to visit the American shores. According to this document, "There was a land on the other side over against their country where people wore white clothes, and carried poles and yelled loudly—and the poles had pieces of white cloth on them." The settlers had heard rumors of this colony in advance and called it "Ireland the Great."[11] Thus, as in the case of the Faroes, Iceland, Greenland, and finally Vinland, the Norse found that the Celtic monks had preceded them.

13 • History and Mystery

A thousand years passed after the first Norsemen stepped ashore on the North American continent before physical evidence could be found to demonstrate that such an event had actually occurred. A period of ten years went into the excavation of the foundations of L'Anse aux Meadows in Newfoundland and the carbon-14 dating of charcoal plus the discovery of household items such as a Norwegian spindle whorl made the claims of the excavators valid.[1] On the other hand, no archaeological find is ever beyond dispute. It would be possible for an individual to claim that the Indians could have laid those foundation stones.

There is always an element of uncertainty in carbon-14 dating, and even though the error were small it might be large enough to throw the whole theory out of balance. Furthermore, some detractor might claim that the spindle whorl was purchased in advance of its discovery, was planted two or three feet under the ground, and then "unexpectedly" discovered several years later as the excavation made progress. Conceivably some person might recall that a yellow-haired woman of a certain age who appeared to resemble Dr. Anne Ingstadt was seen in the company of an attractive man examining spindle whorls in an Oslo curio shop. Two days later, the man returned and bought both of the samples in stock and, when shown pictures of Dr. Ingstadt and friend, the clerks would seem to recall that these were the same customers although the time was eight years before, and who's to say how years change personal characteristics? However, just a whiff of forgery would be enough to discredit this highly significant exploration, and unfortunately these insinuations could never be disproved. In archaeology there is never evidence enough, and unfortunately the greater the mountains of evidence that are heaped up, the easier it is to find apparent discrepancies in the mass of details at hand.

Three additional discoveries that have been attributed to Norse exploits in this country have created a tumult, and it is precisely because these three discoveries are important that a cabal of voices has been raised.

The first of these is known as the Viking Grave at Lake Nipigon. This discovery was made in 1931 by James E. Dodd, who was an employee of the Canadian National Railway. He had settled a claim for some land near the town of Beardmore, about 100 miles northeast of Port Arthur. As Dodd was exploring his claim he found a peculiar slab of dark rock rising ten feet or more above the level of the soil. Through this rock ran a vertical vein of white quartz about a foot wide. Seeing the white stripe of quartz, which apparently pointed into the ground, Dodd seized a shovel and began to dig to see where the quartz would lead. Down three feet, he found a strip of rusty iron which was embedded in the schist, and as he tried to remove the scrap of metal it broke in two. It was not until he had succeeded in extracating the second half that he discovered that it was an old-fashioned sword. He continued to dig and also found an ax, a bowl of sorts which broke into fragments, and finally a piece of rusty iron that looked somewhat like a handle.

At first Dodd thought these had been left by some previous explorer, or perhaps were Indian artifacts, and therefore he tossed them onto a dump heap and left them there for a period of time during which they continued to rust and deteriorate. At a somewhat later time, he decided that he would take them into Port Arthur and see if he could find anyone who would be willing to buy them as curios, and it was then that he met Mr. John Jacob in the Fish and Game Department of the Province of Ontario who was impressed with the possible significance of these iron relics. A report of the finding was made to the Royal Ontario Museum of Archaeology by Jacob, and apparently this institution was uninterested or the mail was never delivered.

Five years later, another representative of the government of Ontario was visiting Port Arthur and heard about Mr. Dodd's find. This visitor was a miner who had a professional interest in the history of metals and he, too, wrote to the Royal Ontario Museum of Archaeology, and this time the mail was delivered.

The museum officials treated these discoveries with appropriate respect, and in their analysis of the findings proceeded with professional care. They not only summoned the best experts on the subject available in this country, they also consulted with experts in the Norwegian Embassy in Ottawa. Their conclusions were clear and unanimous. These implements were unquestionably genuine. They were of Norwegian manufacture. They were made in approximately the eleventh century A.D. The shape, size, and character of the metal all conformed to standards then in existence. There was no doubt about their authenticity.

The news of this discovery leaked out of the museum before any official publications could be issued and received headline attention in Canada, America and in Europe. The following day, one of Mr. Dodd's neighbors sought the attention of the press. He declared that Dodd did not dig these implements out of his excavation as claimed, and that he, the neighbor, had seen these same findings in Dodd's basement long before he ever claimed to have discovered them on his own land. The suggestion was made that a bright young Norwegian scholar and visitor had understood the value of these Norse findings and had brought them from Norway to Port Arthur where they had been sold to Dodd, and Dodd had planted them to be rediscovered at a later date, and that the hoax was an invention of Dodd's mind. The effect of the second story was to cast a shadow of suspicion over the entire discovery. To this day, there is complete agreement on the validity of the scraps of steel that were transmitted to the Archaeological Museum. They were, and are, authentic eleventh-century Norse implements. However, whether Mr. Dodd found them as he explained has been in question every since.

The neighbor in question who challenged the accuracy of the original story later reported that he "was trying to have a little fun with Mr. Dodd, and that he had not really seen the articles in Dodd's basement as he claimed." In view of this second statement, he perjured himself once, if not twice, and his testimony is without validity.

One of Mr. Dodd's partisans has written a touching description of how these arms came to lie in this narrow grave. The old Norseman was traveling with a group of fellow explorers, and whether he died of natural causes or from a wound is unknown. However, his fellow countrymen knew that they were at least 1,300 miles from the eastern seacoast, and therefore the possibility of returning a body for burial at home was impossible. According to this imaginary script, the burial occurred as follows:

> But no one came to claim the body of this first intrepid explorer of American's interior. The way back to Vinland and Greenland was exceedingly far, whether by sea or land, and it is doubtful if any of the company returned to tell the tale. Only of this man can we visualize the funeral up there on the low-lying ridge, sparsely clad with solemn spruce trees. Down to the solid rock they dug his grave, then placed him there beneath the monument that Nature provided. His head was to the west facing the dawn, whence would come the Lord of Resurrection morning. On his left side lay his trusty sword, and on his right his axe. His arms were folded over his breast and over them lay his burnished

shield. Then, perhaps, while his comrades in arms stood around the open grave, one of them repeated as much of the liturgy for the dead as he could remember.[2]

This is a poetic re-enactment of that final interment but unfortunately there are few facts about the Lake Nipigon discovery that justify this reverential treatment. The most important is that there is no skeleton. If perhaps underneath the buckler there had been a rib cage, or beside the sword had been a pelvis, or if the ax had been enfolded in an arm, or gracefully nestled nearby to a skull, the final rites would be more convincing.

Claims are made that a skeleton would in no way survive for a thousand years. Yet we know from excavations around Thjodhild's Church, constructed A.D. 1000, the burial grounds have revealed 150 skeletons of the people who were the first settlers of Greenland. Women are buried to the north of the church, men to the south. It is possible to determine quite accurately the age, the sex, and sometimes the cause of the fatality, and if that ancient warrior had been buried along with his armor, the questions of dating the interment would be relatively simple through the carbon-14 process. Certainly it would settle decisively the question of whether the burial was made after the year 1600 or before the year 1300, and this is all that is required to substantiate Dodd's claim. The fact that there is no skeleton does not mean that these implements were not left there sometime during the eleventh or twelfth century. But on this point, we will have to say that we do not know. We have no way of knowing when the ancient Norse implements were left along Lake Nipigon or who placed them there.

Of all of the architectural anomalies, perhaps none matches that of the Newport Tower in Rhode Island. Scholars of the Norse migration have made great efforts to claim this as one of the original evidences of a Viking colony on the Atlantic coast. The city of Newport, Rhode Island, is very close to the land first sited by Bjorni Herjolfsson. Assuming that Leif Erikson followed Bjorni's track back again to Vinland he might very well have landed at Newport. This would, indeed, have been the land in which wild grapes grew, and it would also have possessed at that time the abundant resources of fish, timber, and wildlife. As the colony grew, the Norse would want to leave some evidence of their early arrival, particularly in view of the fact that the Irish might have a prior claim. They would want to "leave their mark," so to speak, and what better way than to build a tower in the sky.[2] Proponents of the Norse influence in America have made a strong point in claiming that this structure—which is unlike any other building in America—was a gift from the country's early discoverers to future generations.

The building stands in Touro Park in the old portion of Newport, which was the center of seafaring and ocean traffic. The town of Newport was founded in 1639 and the first documentary evidence of the tower was made in 1677 in the will of Benedict Arnold, who had three times been governor of Rhode Island, and who was the grandfather of the superintendent of West Point who betrayed the plans of that fortification to the British during the Reolutionary War. The tower is constructed of mortared stones carefully cut and fitted. It is cylindrical in shape and nine yards high. At one time there was an additional story, but the floor has since deteriorated. It has eight arches supported by round columns with a number of windows and smaller apertures. Authorities on colonial architecture are clear that this building has no prototype in New England.

Doctoral dissertations have been written at Harvard on this subject and surely more will be written in the years to come. No one knows what the purpose of this building was, whether a fortification, a watch tower, a windmill, a church, or some sort of signal station. There is an account of those who believed the building was constructed as a corn mill, and William S. Godfrey excavated the ground around and inside the tower in 1948 and 1949. He found a good many colonial artifacts, including clay pipes and coins, but he found no corn pollen which might permit carbon-14 dating, and efforts to visualize how the building may have been constructed to function as a corn mill have been disappointing and inconclusive. If this design for a corn mill was the brain child of Benedict Arnold, the experiment was a failure because it was never repeated. After all of the theories have been considered and digested, the best judgment of the archaeological world is, in the words of Magnus Magnusson, one of the world's authorities on the Viking's expansion westward, "No on knows what the purpose of this tower was."[3]

In the opinion of this author, the words of Magnusson are the proper point of departure in making a fresh inquiry on who built the Newport Tower. In the long annals of human history, were there any other people who built towers for purposes that no one knew? The answer is yes, and those people were the Celts of Ireland. Before the year 1200, they had constructed more than 2,000 towers in Ireland alone, and more than eighty of them are still standing today. Some of them are as tall as seventy feet, the equivalent of a seven-story building. Like the tower at Newport, some had a number of windows and small apertures, which on casual examination would seem to be quite arbitrary in their placement but which, after careful, systematic study revealed these towers were astronomical observatories. Like Stonehenge in England where giant monoliths have been hauled great distances to be placed on end, the purpose of such construction and the expenditure of such enormous effort appeared to be utterly bewildering.

Not until the design of the modern computer was it possible for man to run time's clock backward and to view again the sun, the moon and the planets as they were seen at the time that Stonehenge was constructed. Not until it was possible to roll back the seasons and calculate accurately the arrival of the solstices and the equinoxes 2,500 years ago could we understand why those enormous stones received their curious placement. Now we know that Stonehenge was a celestial timepiece, that Celts from a great distance must have traveled weary miles to be present at the precise second of midsummer's night's eve, or to celebrate the Celtic New Year which began on the first day of spring. The declination of the earth on its axis, which runs on its own 26,000-year calendar, has shifted the time of arrival of the seasons since Stonehenge was built. But if the careful observer today could look through Newport windows and determine on what day of the year the sun shone through the aperture without making a shadow on either side of the opening, or if the shaft of sunlight from one opening is timed to illuminate a marker on the opposite wall, and if this same observer could calculate back to determine when these celestial events might have occurred, say in the year 1200, we might begin to have some glimpse of the meaning of the Newport Tower.[4]

The third and perhaps most intriguing of the presumed Norse artifacts requires a shift of focus halfway across the continent to Douglas County on the west side of the State of Minnesota, approximately 1,600 miles from the seaboard. Here a farmer by the name of Olaf Ohman, who was a relatively latecomer to Minnesota, purchased a piece of land that had been passed over by earlier arrivals largely because of its rocky contour and marginal fertility. The early Scandinavians had a sense of where the best land was and they got there first. Those who arrived immediately after them did well, but not quite as well. Those who came fifteen to twenty years later took what they could find, and Olaf's farmstead was in this latter category.

In order to improve the land, Ohman had to grub out brush and timber, and on one of these days he pulled up a tree whose roots lovingly embraced a large stone weighing 200 pounds. All evidence showed that the tree had been growing in this location for a minimum of twenty years, and possibly a maximum of forty. As Ohman lifted the tree and continued its removal, he dusted off the stone and found on it curious lettering, which he judged to be an inscription. In this fashion, he not only brought to light an object subsequently called the "Kensington Stone," but also unearthed one of the great debates of archaeology and early American history. The year was 1898, and from that time to the present there have been few writers of early Amerian history who have not had to deal with the Kensington Stone.

After the stone had been cleaned up, the text of the writing was sent to O.J. Breda, a professor of Scandianavian languages at the University of Minnesota. He studied the inscription for several months and made a translation of most of it, although there were several gaps he was unable to translate, and into these interstices later translators made appropriate additions. The final text was agreed to be as follows:

> Eight Swedes and twenty-two Norwegians on an expedition from Vinland to the west. Our camp was by two skerrys one day's journey north from this stone. We went fishing one day. When we came home found ten men red with blood and dead. AVM [Ave Maria] Deliver us from evil. Have ten men by the sea to look after ship's fourteen days' journey from this island. Year 1362.

When Professor Breda was through, he released the statement to the press in which he said that he was convinced that the Kensington Stone was not an authentic thirteenth-century runic inscription. The mixture of Swedes and Norwegians was "contrary to all accounts of the Vinland voyages," and secondly, the language of the inscription was not Norse but was a mixture of Swedish, Norwegian and English. Among the portions of the inscription Breda was unable to translate were the dates, but he guessed that the date of the stone was two hundred years earlier than was later assumed to be.

The Kensington Stone thus started off with a bad press, and the next series of events did nothing to improve the esteem for this alleged historical artifact. After Breda had his shot at the target, the stone was shipped to Professor George O. Curme who was a leading philologist at Northwestern University in Evanston. Curme had photographs taken of the inscription and sent to a number of the leading runic scholars in Europe. None of these men made a detailed study of the writing but returned letters saying that the Kensington Stone was a fraud, and "a clumsy one at that." The Curme Report was now in. Northwestern University decided the stone had no value, or even worse, that some uneducated Minnesota farmer was trying to tweak the nose of the university, and as a consequence the stone was returned to Olaf Ohman with very little thanks and some apparent irritation that the valuable time of the university had been subjected to unbecoming and unscholarly employment.[5] Ohman, who had a sense of personal dignity of his own, was irritated by the condescending tone of the scholars and when the stone was returned he threw it down—fortunately face down—in front of the entrance to his barn where one could walk on it to stay out of the mud and manure. For nine years it lay there. Then a significant event occurred: A prosperous Norwegian farmer by the name of Hjalmar R. Holand came to see Olaf and asked if he could study the stone.[6]

Olaf, who had been subject to unending embarrassment since the time of his accidental discovery, was glad to get rid of the "Lying Stone," and he gave it to Hjalmar Holand, and with this transfer the Kensington Stone became an item of controversy for almost a century.

The scholars who inspected the stone before it was placed in front of the cowbarn to keep the farmer out of the mud had been linguistic specialists. They were the professors of ancient Norse writing at the University of Minnesota, Northwestern University, and European institutions of great distinction. All of them were offended by the literary quality of this document and were unanimous in rejecting it as an authentic runic inscription. During the years ahead, a number of other linguists looked at the stone and came to the same conclusion, and up to the present time all of the challenges to the validity of this stone have been based on the text of the inscription.

At this point, the author would like to make a personal confession that he knows very little about runic writing, but has had a great deal of experience with college professors, and particularly those in the field of literature. I wonder whether any one of the authorities who passed judgment on the Kensington Stone ever had the opportunity to grade freshman English themes as part of his academic experience. The criticisms of the runic text include such items as: verbs are missing, subjects are missing, plural subjects are used with singular verbs, sentences are incomplete, prepositions are misused, words of more than one language are mixed together without any clear understanding of the terms used, words are misspelled or are not words at all, or are inventions of the author. Each mistake listed here is found in almost every batch of freshman composition papers, and presumably these are written by young Americans who have had twelve years of instruction in the English language and were qualified for college discipline. In comparison to the bright-eyed, well-tutored, confident college students, we know very little about the group of thirty men or more who may have composed the Kensington inscription back in the 1300s. We have reason to suspect that they were really quite a bright group of men. In addition to the Norse, Swedish, and English usage, they tossed in a Latin phrase which might establish their Christian credentials. The background of the men who composed this inscription is almost unknown, although we give them credit for courage, commitment, and genuine endurance. Nevertheless, this was the fourteenth century. There were no public schools, and although these were men of above average breeding, it is notable that they could read or write at all. It must be remembered that Cortes, who in 1539 managed to totally destroy the Inca Empire, and who was destined to become a member of the nobility, was illiterate. These men doubtless spoke a colloquial form of the Norse and Swedish languages and not an official version that went into permanent documents. As they traveled together, and for the sake of mutual understanding, they would certainly get the two languages a bit mixed, and were capable of making some substantial errors in the written text.

A number of the critics of this text were almost embarrassed to call it "a runic inscription." I can understand their embarrassment. One leading expert said he detected at least fifty faux pas, and no language expert would ever be able to accept the authenticity of a text with such a bundle of mistakes. However, not one of these scholars would ever be able to produce a "forgery." Their inscriptions would be perfect and thus immediately identifiable. Their writing would contain none of the wonderful surprises of men employed in a task too big for them to understand and for which even the best of them were unprepared.

C.W. Ceram in his book, *The Secret of the Hittites*, comments that the modern scholars who translated the Hittite tablets were required to correct the records of the ancient scribes who slipped a bit in spelling, punctuation, sentence structure and word usage.[7] Contemporary scholars understand the Hittite language better than the scribes who copied down the words of the king and his court; or, to state it differently, these tablets from the royal archives were not good Hittite. Contemporary translators made corrections with indulgence and good humor because they had no occasion to question the authenticity of the tablets which they saw dug out of the ground before their very eyes, and as they found certain oddities and crudities in the text they could afford to adopt a generous and tolerant point of view. The same is true of other languages as well—including English. Today we have professors of Old Norse in the universities who can understand the language of Old Norse better than did the Old Norse. Therefore, it is entirely normal for such men to question the parentage of a bastard script drafted by nondescript sailors, possibly by unscholarly men who might have been illiterate or semiliterate, who were themselves a mixture of language types, and who had picked up a few Latin and English words along the way.

Hjalmar Holand became a burning partisan of the Kensington Stone. He devoted the balance of his life and much of his personal wealth as an advocate of the validity and the authenticity of this unique footnote to American history. As a consequence, he kept the Kensington controversy alive. Since all charges of forgery were based on textual errors, he did not try to challenge the scholars directly, but he did turn up enough knowledge about the Old Norse language that he was able to correctly date the stone and add some new light to the mathematical system of the Vikings. While he was never able to win the support of the philologists or the professional anthropologists in this field, he nevertheless was able to win the support of some historical and antiquarian groups of more than nominal importance.

Holand's approach was to study all of the other aspects that might have led to the journey of exploration and the engraving of the Kensington Stone, and to look at some of the potential events that might have followed the completion of this inscription. He proceeded to do so with the determination of a bulldog and the instinct of a detective. Tenacity and curiosity would normally be considered admirable traits, but they gave offense to the academic community because of the partiality with with he carried on his investigation and the prejudicial manner he had in seeking to win his case. This was contrary to the "spirit of science"that requires the impartial collection of facts and the willingness to accept the conclusion these facts

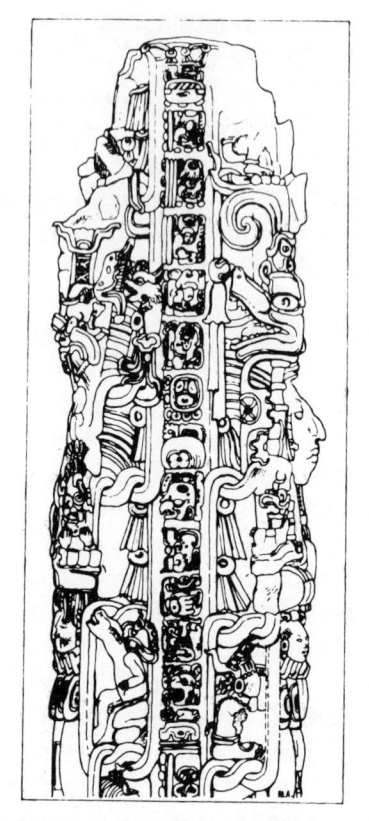

The carved monument pictured here might be called the "Kensington Stone" of Mayan history, although two other stones have turned up with equally unbelievable inscriptions. If this stone were offered as a gift to an unsuspecting museum, it might very well be rejected as a "forgery," and "a clumsy one at that." The reason? Because it contains a horrendous mistake in mathematics that would be unthinkable for any Mayan astronomer who was shrewd enough to calculate the orbit of the earth around the sun for twenty-six millennia in the future. The mistake recorded here is about as intolerable as to add 20 and 20 and achieve a total of 2,020 rather than 40. Yet the mistake is there for all who run to read. To invest so much time and effort in carving and erecting a monument, and to allow an error of such magnitude to slip by all the experts, is inadmissible. Yet the fact is that these mistakes were made by Mayans.

imply. With the exception of the continued defense of Dr. Mudd who removed the bullet from John Wilkes Booth after the assassination of Abraham Lincoln, perhaps no other facet of American history has found a more dogged counsel for the defense than the Kensington Stone found in Hjalmar Holand. In the course of his investigations, he did turn up an enormous amount of information both in this country and from archives overseas on early Norse life and the exploration of the Americas. And although he was partisan in his defense of the stone, it should be said in his behalf that he had the grace to list the names and publications of those who objected to his point of view, and to give a brief statement of their criticisms as well as their sources. In his search for new facts, he managed to keep his critics on a constant defense, and some came to regard him with a certain amount of irritation. Basically, his sin was his failure to respect the authority of the professional. He proceeded on his own to find facts and prove his case. Such a person can be both nettlesome and provocative, and a negative side effect of his efforts is that one important professional group completely were repelled by any endeavor to substantiate the Holand thesis.[8]

There is a positive side, however. In addition to providing the occasion for spilling tons of ink in tabloids, newspapers and professional journals, vast numbers of Americans who might otherwise never be touched by the knowledge of the past have been awakened to some awareness of early American history. Furthermore, anthropology as a field of knowledge has been enriched repeatedly by the very remarkable discovery of amateurs: Troy and Mycenia; the Ventris translation of Cretan Linear B Script; the finding of the Tel El-Amarna archives; and accidental discovery of the Dead Sea scrolls by a shepherd boy—these are only a few of the contributions to the knowledge of human history that amateurs have been able to make. The last word on the Kensington Stone is not in yet, and it will be interesting to see what else time may have to say on this fascinating subject.

Holand had a love affair with the Kensington Stone and in his pursuit of this affection he was able to open archives, persuade bishops to exhume medieval records, and to visit with obscure scholars. Like another fellow countryman, Thor Heyerdahl, he became an experimentalist. He went out and checked the distances in person; he went looking for actual locations where Norse boats might have berthed; and he walked over the alternative routes the crew might have taken for the exploration of inland areas. He went personally to check out the axheads, spears, fire steles, and halberds that were found in the region, and sent them for evaluation to ranking experts overseas. He managed to find the personal diaries of latter-day explorers which offered the prospect of casting light on this earlier if somewhat fabled chapter in American history. He became an irritant to those people who had made up their minds, and his ability to turn up new fragments of information continued to infuriate those people who wanted to let the matter rest.

Holand's thesis for the Kensington Stone was based in part on fact, but portions of it were based on a shrewd insight which was later confirmed by careful research. Some of his ideas were guesswork, but were logical within the framework of reference within which Holand worked. Here in brief is Holand's explanation for the stone which was based on fact, hope, hunch, and careful exploration.

When Bishop Gisle Oddsdon was writing his synoptical history of Christianity in Scandianavia, he made this entry for the year 1342, which read as follows:

> 1342. The inhabitants of Greenland fell voluntarily away from the true faith and the Christian religion, and, after having giving up all good manners and true virtues, turned to the people of America. Some say that Greenland lies very near the western lands of the world.

His reference was to the western colony which is north of the eastern colony of Greenland and was the first to disappear. In general, we assume that a gradual change of climate eventually made life in the northern colony so marginal that perhaps no more than two or three hundred persons remained and it became obvious that they—like their ancestors before them—must move if they were going to survive at all.[9] Regular commerce had been underway with America for three and a half centuries. The members of the eastern colony had a good knowledge of the resources of the country, the coastline, and the opportunities this new land presented. Other sources confirm the accuracy of the bishop's summary. Most of the members of the eastern colony had decided to migrate to America.

At the same time, Magnus Eriksson became the King of Norway and Sweden, and was noted for his missionary zeal to convert the heathen. He had launched one war against the Russians in order to bring them into the fold of the faithful. But this military maneuver was not a success. Magnus appealed to the pope for support to increase the fold and the Pontiff responded positively to Erikson's request for a holy crusade against Russia. In 1351, the pope made good on his promise of support to King Magnus, and he loaned him all of the tithe money collected in Sweden and Norway for 1351 and promised him half of all the tithes collected during the four following years.

King Magnus was now ably funded, but the following year brought the black plague, which not only reduced the population in Magnus's kingdom by 25 percent but spread such death in Russia that it would have been suicidal for an army to march into those lands even for the noblest of causes.

Russia wasn't the only vineyard that required workmen, and when the King became aware of the fact that his own subjects in Greenland had fallen away from religion and had departed for the shores of America, he decided that the funds guaranteed him by the pope should be used for the purpose of bringing his own people back to the fold. He outfitted an expedition and judiciously collected an expeditionary force of both Swedes and Norwegians totaling some thirty men and equipped it with one large merchantman adequately stored for an extended journey, and at a later date added two smaller ships which were primarily for the defense of the major vessel.

This expedition sailed from Norway in 1355, thirteen years after the Greenlanders had given up their holdings in the western colony and migrated to America. The expedition chosen by the King and financed by the pope did not return until 1364, or ten years later. The leader was Paul Knutson, a judge of some distinction who, according to the king's mandate, was instructed to restore Christian worship and, if necessary, to seek the apostates among the people of America whither they had gone; then his task would be either to compel them to return as faithful subjects to the King, or to accept the Christian faith in their new homes.

Arriving in Greenland, Knutson learned what he could about the mass migration to America, and setting out with his thirty men determined to follow and find his fellow countrymen.

There are two alternate theories as to how Knutson found his way to Wisconsin. One is that he entered the St. Lawrence Seaway, continued through the Great Lates, and disembarked some place in the general area of Duluth, which would have brought him within reasonable traveling distance of the Kensington site. However, this route is discounted because of the difficulty Knutson would have had at making the portage at Niagara Falls from Lake Ontario into Lake Erie. However, the possibility of such access into the center of the country ought not to be permanently eliminated.

A more reasonable route for Knutson would be to travel through Hudson Strait into Hudson Bay until he discovered the Nelson River—larger than any in Europe and which obviously drained a continent of enormous size. As did the Hudson's Bay Company after him, he would journey south down the Nelson, in spite of its navigational hazards, to Lake Winnipeg, and then travel south via the Red River which would place him at last on the Western border of Minnesota and not far from the site where the Kensington Stone was found.

Here it was that Knutson split his expedition, leaving ten men to guard the boats and supplies and sending twenty men a distance of fourteen days' travel by foot into the country for purposes of reconnaissance. Apparently, Knutson was following some hunch, or some substantial lead as to where his fellow countrymen had gone; and regardless of how much time it took, he was determined to find them. It was at this point that the massacre occurred when ten men who were on the fishing expedition left another ten men in camp and came back to find them red and dead. The Kensington Stone was a memorial to these ten intrepid explorers who had spent seven years together and who must have endured unspeakable hardship, but had supported each other with good spirit. For those ten men the journey had come to an end.

The inscription was placed on an elevated outcropping of rocks which was once an island in a lake that has since disappeared, leaving the hillock somewhat above the surrounding countryside but far from the most desirable land for tillage and crop production.

Hjalmar Holand carefully scouted out all of these distances. He found the two skerries, or rocky islands that are mentioned in the Kensington text. He found the mooring stones to which the Norsemen and Swedes tied their ships while they set out on their inland expeditions. It was the practice at the time for Norwegian sailors to drill a hole through a large rock which made it possible to tie the prow of the boat to the rock and to anchor the stern away from the shore. Hjalmar found a number of these mooring stones, each with a drilled hole corresponding precisely to those that were used in Norway at the time. It is impossible that Indians had the technology to drill such a stone, and in view of the fact that Columbus had not yet arrived in the Caribbean it is difficult to conceive that other explorers from the east coast might have traveled 1,500 miles inland for the painstaking and time-consuming task of drilling these stones as a safe anchor for their boats.

The Indians thoroughly looted their ten victims. Battle axes and other weapons were carried away by the marauders and have been found over two states but relatively close to the Kensington site. Hjalmar Holand had done a thorough job of identifying these weapons with comparable iron implements manufactured in Norway during the fourteenth century. He has taken wood from the ax handles and sent it for analysis to the Research Center for Paleo-Botany at Harvard University. After study, the curator has reported that the ax handle was made of spruce of a sort grown in Norway. The fact that the wood had mineralized was evidence that the axhead had been submerged in water for several hundred years and thus eliminates the possibility that it could have been deposited in Minnesota during colonial times or any time thereafter. It does not eliminate the possibility that the axhead could have been left by some other pre-Columbian expedition of Norwegian origin, but certainly it could have been equipment of the Paul Knutson exploratory group.

Many of these implements, it should be pointed out, were unearthed in two feet of soil after the first plowing by original settlers. Axes and weapons have been sent to Norway for analysis, and reports have been made which identify the carbon content of the metal, the type of forge on which each was tempered, and even the kind of hammer used to shape the metal. This should settle the claims made by some sociohistorians that these axes were distributed in the nineteenth century as prizes for collectors of tobacco wrappers.

This brings us to the end of the Paul Knutson expedition.

Who or how many returned to Norway in 1365 is unknown. It cannot have been many. Paul Knutson was not among them. What ever became of the Paul Knutson expedition? A few—perhaps a very few—returned to Norway in 1365. In spite of an exhaustive search for the final report on that expedition, it has not yet come to light and must presently be listed as "lost." We do not know how many men returned nor what happened during the ten years that this group of men traveled as personal emissaries of King Magnus of Norway and Sweden. This pope most certainly was interested in what became of his tithe money during those years, and King Magnus as a devoted member of the Church would have written at least an interim report. In spite of the fact that centuries have passed, that report has not been found but may sometime see the light of day. All or part of it may still be lodged in the file of a bishop, a cardinal, or perhaps it has been mislaid in one of the archives of the Vatican. In the meantime, we can review some of the known facts that may cast light on the various alternatives that were open to this group after the massacre at Kensington. A few of them opted to return to Norway, but it is unlikely that they would return to Vinland by walking overland 1,500 miles to New England; rather, they may have chosen to go back the way they came, by water through Hudson Bay.

Another alternative is to assume that most of the members of the expedition determined to continue their search and they were all massacred by the Indians without a single survivor to inscribe another Kensington Stone commemorating this final event.

On the other hand, we could assume that they eventually found the original colony of three hundred Greenlanders and decided to settle among them. Or another possibility is that they may have continued their travels West intending to carry out their original commission but finally found themselves in a set of circumstances that were sufficiently pleasant that they decided to rest a bit from arduous travels, and as they intermarried with the Indian women in this agreeable situation and began to acquire families, what was originally conceived as a temporary stay became a permanent one.

Not far from the Kensington site, there was an unusual settlement of Indians who developed a culture quite unlike that of any of the tribes that surrounded them. The Mandan Indians who occupied permanent villages in the upper Missouri Valley of central North Dakota present one of the fascinating ethno-biological problems of Indians in North America. Unlike the other Indian tribes in the North American region, the Mandans, who lived close to the present city of Minot, were predominantly an agricultural people and their communities were designed and constructed to be permanent residences, unlike the migratory tribes around them who subsisted primarily by hunting. The ethnic origin of these people was mixed. Their skin color ranged from white, or almost white, to the normal brown of the surrounding Indian tribes. The women particularly tended to have fair complexions and to have blue eyes and golden hair, and there were many gradations between these types, including individuals with hazel or gray eyes, and a full range of skin pigmentation combined with eye color. Their religious and social beliefs were not those of the surrounding Indians, and on the basis of the report of the early visitors they appeared to have something in common with Christianity. Their physiognomy was non-Indian in character; their body build was both taller and heavier than that of surrounding tribes.

The first reliable report we have on this community was based on a visit of Captain Pierre de la Verendrye, the first Frenchman known to have entered the region west of the Mississippi in Minnesota. He was at the time involved in an assignment of the French government to find a suitable passage to the Pacific Ocean when, in the course of his travels, he heard that even farther west there were people who were not Indians but were white people like himself. On the basis of such information, he left his exploration to travel several hundred miles to the southwest, and in 1738 found six large villages on the Missouri River. A portion of a report from his pen describing the Mandan village follows:

> M. de la Marque and I walked about to observe the size of their fort and their fortifications. I decided to have the huts counted. It was found that there were 130 of them. All the streets, squares and huts resembled each other. Several of our Frenchmen wandered about; they found the streets and squares very clean, the ramparts very level and broad; the palisades supported on cross-pieces mortised into posts of fifteen feet to twice fifteen feet. There are green skins which are put for sheathing where required, fastened only above in the places needed, as in the bastion there are four at each curtain well flanked. The fort is built on a height in the open prairie with a ditch upwards of fifteen feet deep by fifteen or eighteen feet wide. Their fort can only be gained by steps or posts which can be removed when threatened by an enemy. If all their forts are alike, they may be called impregnable to Indians. Their fortifications are not Indian. This nation is mixed white and black. The women are fairly good-looking, especially the white, many with blond and fair hair. Both men and women of this nation are very industrious; their huts are large and spacious, separated into several apartments by thick planks; nothing is left lying about; all their baggage is in large bags hung on posts; their beds made like tombs surrounded by skins. . . . Their fort is full of caves [caches] in which are stored such articles as grain, food, fat, dressed robes, bearskins. They are well supplied with these; it is the money of the country. . . . The men are stout and tall, generally very active, fairly good-looking, with a good physiognomy. The women have not the Indian physiognomy. The men indulge in a sort of ball play on the squares and ramparts.

Pierre de la Verendrye was impressed with the amplitude of their food, resources, and of their generosity. "They kept on serving feasts without end. They brought me every day more than twenty dishes of wheat, beans, and pumpkins all cooked." Unfortunately, Verendrye lost his interpreter the day after he arrived among the Mandans, and therefore was unable to converse with the unusual people, and as a consequence missed an important opportunity to learn something about their beliefs, history, form of community organization, and customs.

The next record of a visit to the Mandans was made in 1780 by two French missionaries who presented a report on their findings at Montreal in 1783. The following year, the *Pennsylvania Packet* and *Daily Advertiser* ran a story on the Montreal meeting in the issue of August 24, 1784, as follows:

> Letters from Boston mention that a new nation of white people have been discovered about 2,000 miles beyond the Appalachian Mountains. They are said to be acquainted with the principles of the Christian religion and to be exceedingly courteous and civilized. This account was brought by the Indians to Boston, and concurs with others which were reported by two French missionaries at Montreal last year.

In 1832, Mr. George Catlin spent several months living in one of the Mandan villages, and A.P. Maximilian, Prince of WiedNeuwied spent the winter of 1833-1834 in Fort Clark a few hundred feet away. In addition to being an observer, Catlin was also an artist, and with pen and brush he made a collection of sketches which not only give a clear impression of Indian life before the influx of white settlers, but also gives accurate pictorial information about the fortifications, the method of constructing the houses and their family dwellings which were adjacent to a common living room, and portraits of Indian women in their native garb who appear to be very Nordic and beautiful.

In addition to his sketches, George Catlin had the following commentary:

The Mandans are certainly a very interesting and pleasing people in their personal appearance and manners, differing in many respects, both in looks and customs, from all other tribes which I have seen. . . . I have been struck with the peculiar ease and elegance of these people, together with the diversity of complexions, the various colours of their hair and eyes; the singularity of their language, and their peculiar and unaccountable customs, that I am fully convinced that they have sprung from some other origin than that of other North American tribes, or that they are an amalgam of natives with some civilized race. . . .

There are a great many of these people whose complexions appear as light as half breeds; and amongst the women particularly, there are many whose skins are almost white, with the most pleasing symmetry and proportion of features; with hazel, with grey, and with blue eyes—with mildness and sweetness of expression, and excessive modesty of demeanor, which render them exceedingly pleasing and beautiful. . . .

The diversity in the colour of the hair is also equally as great as that in the complexion; for in a numerous group of these people there may be seen every shade and colour of hair that can be seen in our own country, with the exception of red or auburn, which is not to be found. . . .

The reader will at once see, by the above facts, that there is enough upon the faces and heads of these people to stamp them peculiar—when he meets them in the heart of this almost boundless wilderness, presenting such diversities of colour in the complexion and hair; when he knows from what he has seen, and what he has read, that all other primitive tribes known in America, are dark and copper-coloured, with jet black hair.

Both Catlin and Maximilian dwell at length on the religious beliefs of the Mandan Indians, and Maximilian in particular compares them with the other tribes in the surrounding areas. Mandan beliefs differed from other Indian tribes in two important respects: the first is that their original ancestor was the white man, and the second one was that he came from the west in a big canoe. The canoe, itself, became a totem or a part of the hero cult, and in each Mandan community a canoe

208 • *Before Columbus*

No other Indians in North America built lodges like those of the Mandans. Note the high chair in the background reserved for the patriarch and surmounted by a heraldic arrangement of spears and buffalo horns reminiscent of a Viking helmet. In spite of the fact that many similarities exist between this house and dwellings built by Vikings on Greenland and Iceland, it seemed impossible that a house of authentic European design could have been built near the western edge of the Great Lakes. However, in 1964 the foundation of Viking lodges remarkably similar to those of the Mandans was found at L'Anse aux Meadows, Newfoundland, and was constructed by European settlers about A.D. 990 or 300 years before the Mandans built their villages near Minot, North Dakota.

was posted in the central square. No trace of the memory of a white ancestor or the cult of the canoe could be detected in the traditional memories of any of the other Indian communities. Both observers report traditions that seemed to have Christian overtones. The Mandans believe there was once a great flood, with a dove returning with a twig of willow, rather than a olive leaf. There is a tradition of a virgin birth; a story of the feeding of a multitude, which might reflect the New Testament originals.

From George Catlin's pen, we have sketches of the inside of Mandan houses which are unlike those of any other Indian tribe in North America. The structures were round and were supported in the center by a square framework with an open aperture at the top which permitted smoke to escape and sunlight to enter. Around the edges of this large common room were bedrooms or apartments which could be closed off from the common room by means of draperies hung at both ends of the opening. Above the curtained openings of each apartment the man hung his weapons and his more important trophies. Catlin's sketch showed suspended from the ceiling of the room a Viking helmet with two crossed spears. Each one of these residential centers contained a "high chair" or a sort of a throne where the acknowledged leader of that residential group would preside at meetings.

The description of the house, as well as Catlin's sketches made from personal observation, conform in almost every respect to the dwellings of farmers and the rural gentry in Norway during the Middle Ages, except that in Norway such houses were generally covered with bark rather than with hides. The idea of building residential units around a common room was distinctly a Norse invention, and structures of similar design are not only found in literature written during the Middle Ages, but also in reconstruction of Norse homes portrayed in Magnus Magnusson's book on *Viking Explorations Westward*.[10] The high seat reserved for the head of the house is found in Norwegian dwellings as well as in Mandan. Because of the shortage of trees, the American community used turf as a substitute for the wood found in fourteenth-century Norwegian house construction.

How revealing it would be now if we could make some philological study of the language of the Mandans based on our present knowledge of linguistics. How useful it would be to know the blood types of these curious people. But this information we will never have. In 1840 the entire community of the gentle and generous Mandans was exterminated by smallpox.[11]

14 • Problems of Early America

ALL scholars do not believe visitors reached the shores of America before Columbus. Those who reject evidence that visits were made refer to themselves as "isolationists" or "segregationists." On the other hand, the individuals who do believe contact occurred between America and the rest of the world before Columbus are known as "diffusionists." The historical position of the isolationist is that after the disappearance of the land bridge formed at the Bering Strait which physically united Asia and North America until 10,000 B.C., no human contact occurred between the American continents and the rest of the world until the time of Columbus—with one exception. That one exception consisted of the Eskimos who moved from Asia into Alaska about the time of Christ, and who have shuttled back and forth more or less continuously by means of kayak and dog sled visiting family and friends on both sides of the Bering Strait until recently, when intercontinental traffic was terminated decisively by the Soviets. These later-day migrants differed genetically and culturally from the Indians, but they never passed out of the frozen north into the temperate climates.

On the other hand, Americanists who believe that contacts were made before Columbus come in great array and disarray. Some believe that the most seminal contacts came from the East and not from the West. Some of them hold that communication between America and Europe was established only shortly before Columbus, and any documents to the contrary are of questionable validity. Some believe that intervention came from outer space, while others are embarrassed by this apparent sensationalism. Others insist that a variety of exchanges occurred across the centuries and the whole notion of segregation is a myth.

Certain problems of pre-Columbian America have to be dealt with by men of all historical persuasion and among the foremost are these: What is the source of the racial mixture that was evident at various times in Central and South America? Why did a cultural explosion occur among the Indians of Yucatan rather than among older societies with greater capital accumulation, better natural resources, and more defensible territory? What was the process of information collection, institution formulation, and theory building that made it possible for a stone-age people to solve with incredible speed problems of organization and physical mechanics? Was the social structure of Mayan and Incan society based on a family structure or was it based on tribal lines? If new and different social forms existed in these societies, were these aberrations the result of an environmental challenge; were they the consequence of a collection of specialized conditions that will never occur again; or were they the product of highly unique individuals within that society? Is chance the only explanation for the appearance of a constellation of artistic, linguistic, and mechanical inventions on this continent when similar constellations have appeared in other societies elsewhere? All speculation on the history of early America is required to deal in one way or another with these conundrums.

The isolationist answer to the racial mystery is that all of the mixed bag of ethnic types that emerged and were identified in Central and South America came across the Bering Strait as part of the racial stock that settled two continents. Anthropologists are in almost total agreement that Indians of the Americas were migrants of Mongoloid stock who crossed the Bering Strait when the oceans were low and a bridge of land permitted travel by foot from one continent to another.

This bridge may have been open for passage 40,000 years ago, and again from approximately 20,000 years ago to 10,000 B.C. However, the real barrier to passage at all times was not the narrow flow of water that separates one continent from another, but rather the deadly frozen tundra that exists on both sides of the Bering Strait which is impassable and impenetrable, so that very few—perhaps only a handful of families—ever made the transmigration from one continent to another.

According to this concept, all of the racial types which are evident prior to Columbus, and some of which Columbus duly noted, came across the Bering Strait. Perhaps they did not journey at the same time as the original Asiatic stock, but they did arrive as separate groups who were able to retain their racial integrity as they moved gradually from the rim of Alaska down to the plateaus of Yucatan. America was never the home of a single undefiled race, but in its vast spaces was the home of many races who may have been able to exist over the ages with little or no commerce with one another. The blacks whom Columbus thought he noted and who provided the models for the massive basalt sculptures, were not trans-Atlantic passengers from Africa, but direct descendants of the pygmy strain of Negrito still found in the Philippines.

The Semitic types who appear on the friezes and fescos of the Mayans were not Carthagenians but earlier settlers from Cambodia or Loas. Some of these early pioneers apparently came from as far as Australia across a wideness of water in order to traverse by foot a gigantic continent, and thence to risk the Bering passage. This is an awesome journey to contemplate, but it must be held to be within the realm of the conceivable. However, if these primitive people did make the incredible journey attributed to them, and all the time did maintain their racial purity, then some trace of them should be found after they passed over the Bering Strait and began their journey southward to Mezo-America. Here the record is silent. If they built campfires along the way and buried their dead, no trace of either their housekeeping or their ritual has survived, or at least it has thus far escaped our attention. New archaeological discoveries could change all this overnight, but until such evidence comes to light, the proposition that this ethnic mix of Central and South America can be explained on the basis of a multi-racial crossing of the Bering bridge will have to remain a theory, and only a theory.

Evidence of a much more convincing nature exists for a second proposition: that the racial mosaic of Pre-Columbian America is the product of a later-day exchange across the Pacific following the advent of navigation. One solid bit of evidence of trans-Pacific cultural exchange is the sweet potato. Of unmistakable American origin, this diet staple had been transported across the South Pacific as far as New Zealand by the fifth or sixth century A.D. In no conceivable way could this plant have been carried from South American by ocean currents or by the flight of birds, because the reseeding is accomplished only by digging up the tuber and replanting the undeveloped bud under a cover of soil where adequate temperature and moisture assure sprouting will occur. The oceans were an effective barrier to the migration of both plants and animals, and for this reason the American continents developed their unique species. The sweet potato was one plant that could not have been transplanted across the vast spaces of the South Pacific except by the hand of man.

American cotton, which differs genetically from the Afro-Asian variety by having twenty-six chromosomes rather than thirteen, may have been carried by voyagers of the South Pacific from the coast of South America to as far as India. In this case, the possibility exists that cotton could have been borne by birds. However, if birds did convey the seed across the vast expanses of the Pacific, they delayed this transmission for thousands of millennia in which it was theoretically possible for them to do so until almost precisely the time it would also have been possible for a man in a boat to deliver the same seeds by hand.

Thor Hyerdahl demonstrated that a raft of balsa logs could float from Peru to Polynesia, although there is scant evidence that the "long ear" mariners of Peru ever founded settlements beyond Easter Island 2300 miles off the coast of Chile. A voyage of even that length is not without merit, although it shrinks in significance when compared to the vast sweep of 6,000 miles of the South Pacific, from New Zealand to Easter Island, which was negotiated by the Polynesian Islanders themselves no later than the sixth century A.D.

All of the South Pacific Island communities were united by the bond of a common language which continues to the present day. Literally thousands of voyages of exploration and settlement had to be launched—probably because of population pressure—and researchers are convinced that some of these boats, even though they missed every island, inevitably would have reached South America. These explorers of prehistory and pioneers of the sea may have settled on the Pacific coast and intermarried with the Indian stock of Peru and Columbia. Anthropologists of first rank are convinced that indeed they did so. Some of them may have reboarded the South Pacific current to float back home again taking the sweet potato and the possibly American cotton with them.

But these intrepid stone age seamen of Polynesia, with their keelless boats of hollow logs and mulberry sails were not the only explorers of the Pacific. In China and Japan, a literate people with some knowledge of astronomy and mathematics were using metal tools to build carefully designed boats, and some of these navigators were unintentional visitors to American shores. How many Japanese junks drifted to America since the dawn of seamanship is unknown, but in the 19th and 20th centuries alone, sixty instances are recorded of Japanese junks carried off-course into the Pacific, and six of these have landed on the shores of Canada or the United States, and six others have been grounded or shipwrecked off the coast of Mexico. Other unpremeditated voyages like these were made before the advent of the white man—perhaps scores of them in boats of comparable design running back to before the birth of Christ. Indeed, when the first Europeans arrived on the Pacific coast of North America, they found that the Salmon Indians had Japanese slaves who had met their unfortunate destiny as a consequence of the vagaries of the sea.

Accidental intruders found their way to the new world, but intentional expeditions were launched as well. The records of the Shih Chi dynasty report several colonizing efforts before the first century including one under the captaincy of Hsu Fu that consisted of "young men of good birth," representing various skills and trades and "apt virgins!"

The success of these voyages is unrecorded. The colonizing efforts, it should be pointed out, would have had the advantage of following an island-hopping or coast-hugging course all the way to America. Boats could proceed conservatively eastward without being out of sight of land by following the great circle route, and without the need for a carefree abandon of caution that typified the Polynesian who set out on vast voyages to discover minute specks of land.

The physical evidence of human encounter across the Pacific is limited but is of significance. The cultural evidence is more substantial and is growing. Robert Heine-Gelden has made comparisons between Chinese and Mexican art forms that lead him to the conclusion that Asians must have landed in Mexico. Another noted anthropologist, Claude Levi-Strauss, calls attention to an exchange across the South Pacific. He sees striking similarities between Maori tatooing and the body painting of the Cadavco tribe living in Brazil along the Paraguayan border. The style, color, detail, and motif of the body ornamentation of these two peoples are so nearly identical that Levi-Strauss rules out any explanation based on the laws of probability or chance.

Although evidence of travel across the Pacific more than 2,000 years ago is impressive, it does not clarify the problem of racial mix in Central and South America. On the contrary, it complicates and confounds it. The sculptures and the pictorial representation of the Polynesian are not Negroid; they are Polynesian. The oriental influence on Mexican art are not Caucasian or Semitic even though similar elements abound; they are oriental. If anything, the thesis of penetration from the East establishes that America was a melting pot long before Columbus saw these shores.

The Polynesian sailors who out-Vikinged the Vikings were nomads of the sea. In no way could they have contributed to the technical splendor of the Incas who built the empire of Peru. An examination of the mummies of the rulers of Peru reveals no genetic relationship between the Viracocha and the Polynesians any more than it does between the Viracocha and the Indians. No geneticist has yet arisen to assert that the Viracocha evolved from American Indian stock or from the Polynesians. These people of Peru with their unique skeletal and cranial features appeared about 1200 A.D. in the Andean highland almost without warning and their predecessors in South America are unknown. The origin of this racial strain was, and continues to be, one of the fascinating enigmas of pre-Columbian America.

If the tatoo-obsessive Maoris of New Zealand contributed their style of body decoration to the Indians on the border of Paraguay—and the evidence is almost haunting to support the argument that they did—this was a separate cultural transfusion, vastly different from the technology of stone cutting, highway building, irrigation, organization, and diplomacy that typified the Inca action. And if Japanese junks were blown off course to land on the shores of North America—and this seems realistic in light of the evidence that such events continue to occur about once every five years; or if a Chinese flotilla set out deliberately to search the coast until these colonists landed on the shores of Mexico and there influenced the painting of ceramics and the shaping of jugs and jars, still these voyagers from the Orient were not the model for "Uncle Sam" at La Venta, nor the Semitic portraits found in Olmec statuary.

According to the isolationist theory, America remained immaculate until the day of Columbus. Chosen by destiny or by God to be the matchmaker for the American people, Columbus himself was unaware of the unique mission that history had thrust upon him:,to knock first on a virginal door of an undefiled people. It was he who first crossed a threshold, and it was this crossing that permitted the meeting and the mating of two peoples, two cultures, and two worlds.

The segregationists of American history—and there are many—have defended this integrity of the native American with a scrupulosity that is religious in its intensity and is commendable in its thoroughness. For historians of this persuasion, the story of Quetzalcoatl is not a dissimulation as much as it is a fairy story—a charming product of the Indian mind. This white man with yellow whiskers was a chimera efflorescing in a vacuum. The Indians, like all other people, had their dreams and their nightmares. Like all people, they had visions of things that never were. The Greeks dreamed of centaurs, mermaids, and satyrs. The European of an earlier age could accurately describe a unicorn. The Chinese dreamed of dragons. The Celts had conversations with hobgoblins, leprechans, and witches. The Norse conjured up visions of Valkyrie carrying fallen warriors from the battlefield to heaven. And the American Indians dreamed of a white man with a yellow beard.

The great tragedy of the Aztec was that his mythology was used to destroy him. The yellow-bearded white man is unquestionably part of the Indian pantheon. Nevertheless, critics who insist that Quetzalcoatl was a figment rather than a fact point out that in some cases he is portrayed as having a black face and a yellow nose. While he objected to human sacrifice, he is at least guilty by association of being portrayed with priests and gods who did practice sacrifices. Thus, the enquirer is confronted with ambiguities which raise a serious question as to whether there actually was a fair-haired man who visited the Caribbean shores before Columbus, or whether the entire story is a fabrication.

In assessing the merit of the account, one should recognize that Quetzalcoatl appeared as a critic—not only of the ruling priesthood but of the morality of everyday life. In this role of prophet, he emerges as a believable person who is unable to protect himself from his enemies, nor to shield himself from the indignities heaped upon him. He used no magic and claimed no supernatural powers. While his preachments on honesty and fair dealing were obviously irritating, he was sought out by others and at least for a time enjoyed a degree of popularity. He taught both fidelity and moderation, but he loved to dance and enjoyed the pleasures of the gregarious. Had he lived in the land of grapes, he was a man who would have been asked to make wine at weddings. Finally, he lost the gamble to point out corruption and call attention to the malpractices of the powerful. He suffered the fate that men of his ilk have known over the ages.

It is not necessary for a society to invent such a man. He is part of the record. He exists in all societies and in all generations, from the dawn of history to the "enemies list" of the White House. The fate of he who criticizes power is known. Why the Indians, who were capable of unbridled flights of fancy and of splendid voyages into the fantastic, would wish or need to fabricate such a personality is a monumental puzzle.

Finally, the student of early America must face the question of the emergence on these shores of certain parallels of style and form that are identical to or closely related to forms and conventions in other societies. Do such concurrences happen as a matter of chance? Is there a law of probability or some mathematical basis for determining how often a convention occurs and is adopted? In the history of measurement, for example, the variety of units for calculating linear distance are legion and vary with each society and with each generation that devises them. Some units for the measurement of distances include the foot, the hand, the span, the finger, the cable, the stride, the meter, and the cubit. The engineers and the architects of the Mayan empire used a unit of measurement precisely the same as the Egyptian cubit. Could this have been a coincidence?

Long ago in the Egyptian society, the circle was divided into 360 degrees. Why 360 degrees? Why not 100 degrees? Or 400 degrees, or 500 degrees, or 1,000 degrees? We have no explanation today as to why the figure 360 was chosen. Certainly today we feel a certain attraction for a system based on units of ten, and for the Greeks and Romans who were never very good at decimals or fractions, a circle of 400 degrees would have had more workability in terms of extending whole numbers than one of 360 degrees.

Someone in ancient time also divided each degree in a circle into sixty minutes and each minute into sixty seconds. At about the same time, someone cleverly divided the hour into sixty minutes and the minute into sixty seconds so that the movement of the sun across the sky became a measurement of time as well as of space. It was precisely this unit that was used in building the pyramid of Cheops and the pyramid of the Sun in Mexico.

Was the discovery of this unit of measurement a fortuity, a matter of chance? Nigel Davies in *Voyagers to the New World* finds an explanation for remarkable coincidences of this sort that is not based on cultural transmission, nor on the mathematics of probability. According to Davies, these coincidences are products "of a common heritage of the human mind, derived from the dreams of an unfathomable past." Certain archetypal images, deriving from prehistoric religion and ritual are inherent to the human psyche, just as animals have built into their nervous systems certain instinctive patterns of fear and flight which are transmitted from generation to generation by genetic means. According to theory, it is from this inheritance that we find an explanation for the recurrence of common forms and conventions in societies which are apparently separated by time and distance.[1]

As attractive as this theory of human behavior is—and it has had the support of some of the intellectual giants of the modern age, including Darwin, Freud, and Jung—countless experiments by psychologists and physiologists over the past fifty years have failed to demonstrate that such an inherited archetypal memory exists, or if it does exist that it influences our behavior today.

No more eloquent or profound spokesman for the isolationist position can be found than Davies who states with conviction: "Science, however, has failed—with one exception —to reveal a single trace of other visitors to the New World during the millennia that separate Columbus from the submerging of the Bering gateway."[2] This brings us to the fundamental questions of "What is a trace? What is evidence?"

Recently a gold coin was found in Venezuela on the shore of the Caribbean. Presumably, this coin was minted by the Greeks in Egypt at the time of Alexander. Of what is this evidence? The answer is nothing, because we have nothing to relate it to. But perhaps sometime we will have knowledge enough to understand what this tantalizing bit of evidence signifies.

Until recently none but a few believed there was any trace of a Viking presence in America before Columbus. Hjalmar Holand was one of these few, and he made a nuisance of himself by turning up seemingly useless and misguided cues: the axe heads he turned up after first plowings were proven to be prizes given for collecting tobacco wrappers; the swords and spears he found were genuine 11th century Viking weapons, but they were buried here by our own contemporaries as a practical joke; the mooring holes drilled in rocks by Viking sailors as anchors for boats certainly looked like the same kind of a hole drilled in Norway stone—but who can tell anything from a hole in a rock? The crystalized wood from an iron tool at the bottom of a lake was 600 years old, but who knows how that tool found its way to the bottom of the lake?

The L'Anse aux Meadows provided respectable scientific evidence of a Viking settlement in America in 989. Suddenly the findings of Hjalmar Holand moved from the plausible to the probable. Overnight, critics realized we had now turned up hundreds of traces of the Norse presence in America. Enlightened by our knowledge of the Norse penetration of Europe and the east, we were able to see in a new light the Norse exploration of America and the west. We knew the Vikings had captured England, stormed Paris, fought the Russians, plied the Volga, sailed the Caspian, traded with Constantinople, and followed routes over sea and land more than 5000 miles from their homeland. In terms of what we know of them, it is now possible and even probable that they explored the entire east coast of North America, penetrated the continent from Hudson Bay, camped on the shores of the Great Lakes, and developed a workable knowledge of the surrounding territory.

In 1977, a dime-size piece of metal was unearthed in an Indian mound near Round Rock, Texas. This coin was minted in London in 313-314 A.D. and is found so frequently in Europe that it is not worth faking. On one side, the coin bears the likeness of a young Roman Emperor wearing a laurel crown, and on the reverse side it is inscribed with the Sun. The coin was found in an Indian mound with other artifacts that make dating possible at about 800 A.D. It is one of fifteen similar coins that have been found in the United States, many of them discovered under comparable circumstances. Is this "evidence" of a pre-Columbian visit?

Not according to Jerry Epstein, anthropologist, who theorizes that wrecked ships could have drifted across the ocean carrying a treasury of coins. In other words, the ship made the passage, but the sailors didn't. "I can picture

Indians gathering coins from the ship," the anthropologist said, "and passing them around, passing them from one person to another in trade."[3]

Another theory is that early coin collectors went visiting Indian mounds and lost their prized possession—not once but fifteen times. "There probably were such coin collectors in the U.S. even in Washington's day," according to Epstein.

"Some early settlers may have dropped it accidentally, or given it to the Indians."

Early American hobbyists presumably would take their coin collections with them when they went to inspect Indian mounds. If this strikes one as odd, one wonders why they persisted in losing the same coin, or why they gave these curios to unlettered Indians when only the donor had the faintest inkling of the intrinsic value of these ancient Roman coins. What strange compulsion drove these early collectors to Round Rock, Texas, there to engage in this bizarre behavior also remains unexplained. But actually, these coins may one day be regarded as evidence of something else; as solid proof of a previously undocumented cultural transaction in American history.

Readers of the Soviet press were informed on May 25, 1979, of a discovery that had been confirmed by Mexican anthropologists.[4] Amber ornaments had been found in the sepulcher of a Mayan Indian tribe who lived in the eighth century B.C. As is generally known, the shores of the Baltic Sea are the native land of this stone. Along with tin and purple die, amber was a special feature of Phoenician trade and through strict secrecy, ship captains maintained a monopoly on this gem that was admired by the Greeks and the Egyptians. What this new find conveys is not yet certain, but it may be an additional clue to confirm the correctness of the supposition of contacts between Europe and America long before it was discovered by Columbus.

Notes

Chapter 1: The Testimony of Columbus

1. Arthur Morgan, *Nowhere Was Somewhere* (Chapel Hill, N.C.: University of North Carolina Press, 1946), p. 198.
2. Ibid., p. 197. See also Peter Tompkins, *Mysteries of the Mexican Pyramids* (New York: Harper and Row, 1976).
3. Leo Weiner, *Africa and the Discovery of America*, (Philadelphia: Innes and Sons, 1920), 1:34.
4. Samuel Eliot Morison, *The European Discovery of America* (New York: Oxford University Press, 1971), 2:5.
5. Ibid., p. 144. One wonders why Don Juan would give Columbus such a tip and inform him precisely of the best passage from Africa to South America—information that the Portuguese king could gather only after considerable expense, and the trial and error of previous voyages.

Apparently Columbus had some of the same doubts as he followed the route Don Juan had predicted would lead him to the continent. Why was Juan so sure? Could he have been concerned with Columbus's welfare deeply enough that he would give away so priceless a secret? Was it a trap? Perhaps these or other thoughts were in Columbus's mind on August 4 when he abruptly changed his course only twenty miles short of the coast of South America. After turning back, he redirected his fleet to the north where he knew there was a certain landfall, a maneuver that was highly unusual for Columbus.

As a result of this change of course, he discovered the island of Trinidad as he sailed along the north shore of South America, and delayed for three days his landing on the true continent. On August 5, 1498, Columbus first set foot on the mainland. When he was satisfied that he had discovered a continent, he wrote in his log that he had found an "Other World" rather than a "New World." This choice of phrase cost him the distinction of having two continents named in his honor.

6. Ibid., 1:32-80.
7. Weiner, *Africa and the Discovery of America*, p. 198.
8. Morgan, *Nowhere Was Somewhere*, p. 200.

Chapter 2: The Celtic Essence

1. Barry Fell, *America B.C.* (New York: Quadrangle/The New York Times Book Co., 1976), p. 80-92.
2. Ibid., p. 96-111

3. Anne Ross, *Pagan Celtic Britain* (London: Sphere Books, Ltd., 1974), pp. 105, 107, 116, 127-128, and others. Also see ibid., pp. 217-31 for illustrations of phalic sculpture found in this country, and comparable examples from Iberia, Brittany. Also illustrated examples of phalic sculpture and Ogam writing.
4. Peter Tompkins, *Mysteries of the Mexican Pyramids* (New York: Harper and Row, 1976), p. 131.
5. Fell, *America B.C.*, pp. 111, 115, 162.
6. Ibid., p. 230.
7. Ross, *Pagan Celtic Britain*, p. 48.
8. Fell, *America B.C.*, pp. 106-7.
9. Ibid., p. 9.
10. Ibid., pp. 26-37.
11. Ibid., p. 64.
12. Ibid., pp. 159, 224-75.
13. Merle Severy, "The Celts," *National Geographic*, May 1977, pp. 582-633.
14. Ibid.
15. Ross, *Pagan Celtic Britain*, pp. 45-64
16. Severy, "The Celts."
17. Ibid.; see also Ross, *Pagan Celtic Britain*, pp. 94-172.
18. Severy, "The Celts."
19. Ibid.
20. Gerhard Herm, *The Celts* (New York: St. Martins Press, 1977), pp. 274-81.
21. It would be unfair to these remarkable people to leave the impression that the Celtic contribution was confined to the realm of the aesthetic and the spiritual. These indestructible individualists were also able inventors, and a short list of their credits can only be suggested here. They were the first to make soap and who today can dream of a civilized society without it. Chain armor was their invention and they were the first to learn how to shoe horses. They fashioned the first files and chisels. It was they who first conceived of a seamless iron rim for the wheels of their chariots and wagons. On the basis of their experimentation, they established standard guage at 4 ft. 8½ in. still used by railroads and in the manufacture of cars. They developed the handsaw and pioneered the iron plow. They are credited with being first to use a number of agricultural implements which significantly increased the productivity of human labor. In addition to material inventions, they devised important social institutions, such as the first parliament, trial by jury, and women's rights.
22. Ibid., pp. 43-44.
23. Roland H. Bainton, *The History of Christianity* (New York: American Heritage, 1964), pp. 136-137.
24. Ibid., p. 104.
25. Herm, *The Celts*, pp. 256-73.
26. Bainton, *The History of Christianity*, pp. 145-46.

Chapter 3: Celtic Migration

1. F. E. Warren, *The Liturgy and Ritual of the Celtic Church* (Oxford: Claredon Press, 1881). The Celtic church was based on the traditions of Paul rather than Peter. Some members of these Celtic communities were "lay ministers" a "priesthood of all believers." Some were married. St. Patrick refers to himself as "the grandson of a Priest." The Celtic Christian church survived in Cornwall until 936. The Celtic church in Wales was not Romanized until the twelfth century. Celtic Christian churches survived in Scotland and Ireland until the time of Bishop Herbert (A.D. 1147-1164). Today the Celtic spirit survives although the churches are Roman. Pre-Christian practices also survive in the folkways of these people.

Monasteries varied in size. Those at Clonard and Bangor each had more than 3,000 residents. The known crafts and vocations were all represented. They contained schools, seminaries and publishing houses (scriptorial) and were depositories for the classical culture of Greece and Rome.

2. Estyn Evans, *Prehistoric and Early Christian Ireland* (New York: Barnes and Noble, 1966).

3. Brian DeBreffny, *The Churches and Abbeys of Ireland* (New York: W.W. Norton & Co., 1976). The evidence of pre-Christian Celtic influence is found in many of the surviving structures, including the "Cult of the Head," e.g., head with no neck.

4. *Encyclopaedia Britannica*, Vol. 1, 14th Edition, see Alaric, p. 497.

5. Ibid.

6. Aziz S. Atiya, *History of Eastern Christianity* (London: Methuen, 1968).

7. Roland H. Bainton, *The History of Christianity* (New York: American Heritage) p. 144.

8. Merle Severy, "The Celts," *National Geographic*, May 1977.

9. *Encyclopaedia Britannica*, 14th Edition, see Patrick. Many scholars consider the St. Patrick story a fabrication manufactured by Church fathers 100 to 400 years after the event. Like other Irish legends, the authentic elements in this story may have undergone imaginative evolution.

10. Bainton, *History of Christianity*, p. 137.

11. Samuel Eliot Morison, *The European Discovery of America* (New York: Oxford University Press, 1971), The Northern Voyages p. 13-31. No artifacts have been found on this side of the Atlantic that specifically identify the landing of the Brendon expedition. Morison is inclined to dismiss the voyage of St. Brendon as another Irish fairy story, along with the legend of King Arthur, etc. However, as much as the Irish may have embroidered these accounts of their early history, many scholars feel that by comparing varying texts there is a common body of fact that underlies them all. While no evidence of Brendon's landing has thus far been discovered in the Americas, other Celtic artifacts from this period apparently have been found.

12. Barry Fell, *America B.C.* (New York: Quadrangle/The New York Times Book Co., 1976), pp. 159, 224-75.

13. Tim Severin, *The Brendon Voyage* (McGraw-Hill, New York, 1978). Contemporary Irish lads replicate the voyage of the beloved saint using similar materials and

methods of boat construction as those used in A.D. 484. They crossed the Atlantic the hard way, however, by following the island-hopping pattern attributed to Brendon which took them more than a thousand miles out of the way via Iceland, Greenland, and Labrador through the most violent seas at a hazardous time of the year. Following a less arduous route, solo voyages across the Atlantic are made almost every year in craft less than a fourth the size of Severin's, and trans-Atlantic races of one-man crews are conducted on a biennial basis. Contains a quick survey of the literature on the Brendon saga.

The Voyage of the Brendon "had proved beyond doubt that the Irish monks could have sailed their leather boats to the New World before the Norsemen, and long before Columbus."

"In proving that such a long-ago voyage could have been made, Tim Severin and his crew have brought one of history's most intriguing tales a giant step closer to the realm of possibility," said the editor of the *National Geographic* in "Who Discovered America? A New Look at an Old Question," December 1977.

Chapter 4: Tools for Discovery

1. Peter Tompkins, *Mysteries of the Mexican Pyramids*, (New York: Harper and Row, 1976), p. 191.
2. Earl H. Swanson et al., *The New World*, (Oxford: Elsevier-Phaidon, 1975), p. 46.
3. Ibid., p. 46.
4. Ibid., pp. 56-57.
5. Ibid., p. 55.
6. Ibid., p. 10-11.
7. Thor Heyerdahl, *Aku-Aku*, (Chicago: Rand McNally, 1958), pp. 347-48.
8. Magnus Magnusson, *Viking Expansion Westwards*, (London: The Bodley Head Ltd., 1973), p. 114.
9. Swanson et al., *The New World*, pp. 51-52.
10. Barry Fell also cites studies of potsherds found near Celtic ruins as evidence of his claim that migrations arrived from Europe about A.D. 200, pp. 93-100.
11. Swanson, *The New World*, p. 57.
12. Thor Heyerdahl, *The RA Expeditions*, (New York: Doubleday, 1972).

Chapter 5: Utopian Mystery

1. John Collier, *Indians of the Americas* (New York: Norton, 1952).
2. Sir Thomas More, *Utopia* (New York: Washington Square Press, 1965). This book was first published in Latin by Erasmus and other friends in Flanders.

3. Thor Heyerdahl, *The RA Expeditions* (New York: Doubleday, 1972), pp. 37, 39, 41-47.

4. Arthur Morgan, *Nowhere Was Somewhere* (Chapel Hill: University of North Carolina Press, 1946), p. 197.

5. Ibid. Morgan's book contains a reproduction of this map.

6. Ibid., p. 195.

7. W. H. Prescott, *Conquest of Peru* (New York: Hurst & Co., 1847).

8. Earl H. Swanson, *The New World* (Oxford: Elsevier-Phaidon, 1975), p. 30.

9. Morgan, *Nowhere Was Somewhere*, p. 46.

10. Ibid., pp. 46-47.

11. Including the Inca as head of government, according to Prescott, *Conquest of Peru*, pp. 79-80, "The Inca himself did not disdain to set the example. On one of the great annual festivals, he proceeded to the environs of Cuzco, attended by his court, turned up the earth with a golden plough—or an instrument that served as such—thus consecrating the occupation of husbandman as one worthy to be followed by the children of the Sun."

12. More and Prescott describe an almost identical work day. See Morgan, *Nowhere Was Somewhere*, p. 50.

13. Minstrelsy, balladry, poetry reading, and play production were probably regarded principally as entertainment. But like the classic Greeks, they were a major force for adult education. Plays were either comedy or tragedy; sometimes a mixture. Epic poems kept alive the memory of the great events and persons of the past. Traveling groups conveyed information not only about recent events in the empire, but also stimulated an understanding of the religious and philosophical basis of Incan life. They played an important role in cementing the various cultures of the empire. Parts were learned by rote. No text survives. Prescott, *Conquest of Peru*, pp. 75-76 et seq.

14. Ibid., pp. 79-82.

15. Morgan, *Nowhere Was Somewhere*, pp. 47-48.

16. Prescott, *Conquest of Peru*, pp. 34-35.

17. Ibid., p. 33.

18. Morgan, *Nowhere Was Somewhere*, p. 21.

19. By this process a form of democratization began to develop and a type of nonheriditary nobility was created. The size of the civil service must have been larger than comparable European states at that time. A "merit system" was devised to operate the bureaucracy of the Empire and a special educational programs was designed to train public administrators.

20. The Incas believed in "one God," but it was a complex theology which also incorporated the idea of a "Trinity" and therefore should be understandable to Christians. According to Prescott, op. cit., "The Peruvians, like so many other Indian races, acknowledged a Supreme Being, the Creator and Ruler of the Universe whom they adored under the different names Pachacmac and Viracocha. No temple was raised to this invisible spirit, save one only. . .," p. 55.

Second, "The deity whose worship they especially inculcated and which they never failed to establish wherever their banners were known to penetrate was the Sun. It was he who presided over the destiny of man," p. 56.
The third element in this Trinity was the Inca himself. He was both son of man and son of God. He was "son of the Sun"; a "child of the Sun." His body was mortal, but destined for resurrection. His soul was immortal. All believers in the Sun, and all Incas, were "children of the Sun." All descendents of the original Inca with white skin and red hair were "Viracocha"—a term meaning "Foam of the sea"—an argument, according to Prescott, "for deriving the Peruvian civilization from some voyager from the Old World," p. 55.

21. Ibid., pp. 26-27.

22. Ibid., p. 15.

23. The protection of animals was part of a "gentleness ethic" that permeated the social structure.

24. Prescott, *Conquet of Peru*, p. 98. See also pp. 42-45.

25. The Incas developed a process of mummification that was as effective as that of the Egyptians. The deceased bodies of the ruler and members of the royal family were carried into the highest mountains where extreme cold and extreme dryness dehumidified the bodies. The mummies of deceased rulers were ensconsed in the Temple of the Sun at Cuzco where, on festive occasions, they were brought into public view to preside over banquets and ceremonies. At one time all of the mummies of deceased rulers were kept in respectful storage at Cuzco. Ibid., pp. 59 et seq.

26. From the very first contact the conquerors were struck by the difference in the appearance of the Viracocha and their subjects—the Indian race. The process of maintaining these differences was preserved by a very careful process of regulated marriages described by Prescott. Studies of the difference in cranial size and shape of the Incas in comparison to native Indian tribes go back to 1750 and are found in Spanish archives. A Philadelphian by the name of Dr. Morton published a treatise in 1825 giving sketches and measurements showing that the cranial cavities of the Viracocha were considerably larger than those of the Indians. For the latest findings on the relationship of brain size to intelligence see the writings of the eminent scientist, Carl Sagan, in *The Dragons of Eden. Speculation of the Emergence of Human Intelligence* (New York: Ballantine Books, 1978).

27. Cecil Howard, *Pizzaro, and the Conquest of Peru* (New York: Harper and Row, 1968), pp. 11-25. The editors of this volume returned to the original Spanish sources, but added little new insight to the still authoratative work of William H. Prescott.

28. Ibid., pp. 50-56.

29. Ibid., p. 75.

30. The similarities between Christianity, particularly medieval Christianity, and the Inca way of life were noted by members of Pizarro's party and became the special interest of subsequent Spanish clerics, judges and administrators. Prescott returns to this theme repeatedly in his *Conquest of Peru*. He notes similarities in the ceremony for the investment of young men in the Incan nobility with the rites of investiture of a

medieval Christian kinght. In both cases they were clothed in a white toga wearing a red cross on their breast, and each candidate was dressed by the high priest in person and shod with white leather sandals.

The use of bread and wine at communion appeared to be symbolized at the religious festivals held each month, and at the great festival of Raymi held each year. The bread was made of a specially ground corn flour, and the wine was prepared from the fermented pulp of the corn stalks. pp. 65-66.

In addition to communion, the orthodox Spanish observer seemed to see a striking resemblance in the practice of confession and of penance. p. 66.

"Another singular analogy with Roman Catholic institutions is presented by the Virgins of the Sun, the "elect," as they were called, to whom I have had occasion to refer," says Prescott on p. 67.

"The authority of the Inca might be compared with that of the Pope in the day of his might, when Christendom trembled at the thunders of the Vatican, and the successor of St. Peter set his foot on the necks of Princes," p. 100.

The character of the god Viracocha as part of the Incan Trinity has been noted earlier, and the similarity between this diety and the Jewish Yahweh are many. No temple was raised to this invisible being, save one in the spiritual capital, and this had its "Holy of Holies" which could be entered only by the Inca himself. No graven images were permitted, nor could portraits be made of a god who was creator and universal spirit. He could, indeed, be worshiped only "in spirit and in truth," p. 55.

"The mild and docile character of the Peruvians would have well fitted them to receive the teachings of Christianity, had the love of conversion, instead of gold, animated the breasts of the Conquerors," p. 103.

Chapter 6: Utopia: Land of Prisoners or Persons

1. Arthur E. Morgan, *Nowhere Was Somewhere* (Chapel Hill: University of North Carolina Press, 1946), pp. 23-32.

2. William H. Prescott, *The Conquest of Peru* (New York: Hurst & Co., 1847), p. 85.

3. Morgan, *Nowhere Was Somewhere*, pp. 149.

4. Arnold Toeffler, *Future Shock* (New York: Random House, 1970).

Chapter 7: The Skills of the Phoenicians

1. No one has ever seen a Phoenician ship, however, excavation now under way may give us something to consider. Miss Iris Love, who is exploring the ruins of the City of Cnidus in southern Turkey, believes that the representation of five sunken triremes can be seen in the silted harbor of Cnidus. If these are ships described by Herodotus they will be of Spartan construction. However, until the excavation is carried out, nothing is certain. *The New Yorker*, June 17, 1978, p. 46.

2. Homer, *The Odyssey*, trans. by Albert Cook (New York: The Norton Library. 1967), pp. 556-60.

3. Ibid., pp. 562-65.
4. Samuel Eliot Morison, *The European Discovery of America* (New York: Oxford University Press, 1971), p. 5. See also Peter Tompkins, *Mysteries of the Mexican Pyramids* (New York: Harper and Row, 1976), pp. 348-51.
5. Thor Heyerdahl, *The RA Expeditions* (New York: Doubleday, 1971).
6. A few rare bronze artifacts were discovered among the possessions of the Incas. While tin is found in Bolivia, the paucity of bronze artifacts suggests that the complicated five-step process of refining this metal had not yet been developed, and the tin for the manufacture of those few bronze items, or the actual implements themselves, may have found their way to Peru from European sources.
7. *Oxford Dictionary of Quotations* (New York: Oxford University Press, 1955), see passages on Ulysses by Alfred Lord Tennyson.
8. Tompkins, *Mexican Pyramids*, 352.
9. Ibid., pp. 348-51. It is a matter of some luck that we have any record of this voyage. A narrative of this mission was posted on the temple of Baal in Carthage as a thank-offering. A visiting Greek was intrigued by what he found and copied it down, attributing the authorship to Hanno himself. The only text that survives is the version in Greek. Scholars are uncertain as to the precise year in which the expedition occurred and assume that it may have been as early as 520 B.C. and as late as 470 B.C.

There is no evidence from the Greek version that any of Hanno's ships were blown off-course, or that any of his colonists ever reached the Carribean. A colonizing expedition of similar size and strength was launched at about the same time under the Captaincy of Himilco, and according to Pliny this fleet of colonists sailed West and North into the Atlantic. Fridtjof Nansen believed that at least part of this expedition reached the coast of Britain.

The first Phoenician settlement on the Atlantic coast outside of the Straits of Gibraltar was the city of Gades (now Cadiz, Spain) founded in the Twelfth Century B.C. By the time of Hanno's voyage, Carthage had established 300 colonies along the Atlantic coast of Africa for more than 2600 miles south of the Straits of Gibralter. These settlements stretched from Algiers to the present-day Sierra Leon. Eight hundred years later, Saint Augustine who was the Bishop of Hippo in Africa declared that in his day (A.D. 430) the Phoenician language was spoken all along the Atlantic coast of Africa as far south as Cape Noun.

The fact that many colonizing voyages had been and were being made into the Atlantic over a span of 800 years, and that some of them were massive enough to establish settlements of considerable size is more relevant than the fact that we cannot detail that any of Hanno's boats reached the Yucatan Peninsula. What is significant about Hanno's record is that he sailed south beyond the "bulge of Africa." Coming home, he would battle powerful headwinds and currents along the coast that would make it virtually impossible for his fleet to round the bulge without sailing far to the West into the North Equatorial current as mariners have done both before and since. Should any ship on this return voyage have met with a sudden storm, "they would have been thrust irresistibly toward the Antilles," and the coast of Central America, according to Armandeo Cortesao, a leading Portuguese oceanographer.

10. Morison, *European Discovery of America*, p. 96.
11. Ibid., p. 12.
13. With the exception of the island-hopping route of the Vikings, the shortest distance across the Atlantic is 1,020 miles from Corvo to Newfoundland, or 1,540 miles from Africa to Brazil. The Phoenicians or the Egyptians may have been the first to make this daring journey, but whoever they were, credit for an even more breathtaking accomplishment goes to the Polynesians, who sailed to Hawaii at a date no later than A.D. 750, and possibly as early as A.D. 500.

The ocean journey of these intrepid sailors was not less than 2,300 miles and may have been longer. These stone-age people who launched an incredible voyage of discovery sailed in ocean-going canoes hollowed with stone adzes and with boats lashed together to form a catamaran of greater stability. With no knowledge of weaving, they made sails from hides or from the pith of mulberry stems. They had no written language, no system of mathematical notation, and no instrumentation. If the first voyage was launched from the Marquesas Islands, as now seems likely, it was necessary to cross three powerful ocean currents and counter-currents, each one of which would forcefully thrust them out of their way, and to pass through an ocean wilderness that generates somes of the most unforgiving storms of the South Pacific.

By contrast, the first trans-navigators of the Atlantic set off on the predictable north equatorial current and would inevitably reach shore if they continued to sail west. The voyagers from the Marquesas, on the other hand, set out for a speck of land they might easily miss, and it is possible that many canoes were launched before a discovery was made. In terms of probability theory, a mariner setting off from the Marquesas to discover Hawaii had perhaps one ten-thousandth of a chance for a successful landing in comparison to boatsmen putting off from Morocco to find the new world. However, once the Hawaiian Islands were located, regular passages back and forth from the home bases were established and continued until A.D. 1275, when they suddenly broke off because of a religious controversy. Hawaii, the las. state to be admitted, may have been the first to be discovered and colonized, and the Polynesians who did it had a flare for enterprise and daring that would be a credit to the nation of which they were destined to become a part.

Chapter 8: The Wisdom of the Egyptians

1. Thor Heyerdahl, *The RA Expedition* (New York: Doubleday, 1971), pp. 33-47.
2. Hatsheput's burial temple is one of the choice gems of Egyptian architecture. It was constructed on the west bank of the Nile opposite Luxor and was destroyed immediately after Hatsheput's death by her nephew, Thutmose III, when he ascended the throne as Pharaoh. He instructed that all the stones of the temple be broken into pieces and buried in deep wells where they remained, and all records of this remarkable structure were lost. The pieces were discovered by Polish archeologists where they were hidden by Thutmose III's pillagers, after World War II. This exquisite but spacious structure is still being painstakingly reassembled but is already one of the most nearly perfect and complete survivals of Egyptian antiquities. For a

sensitive treatment of the life and loves of Hatsheput see Barbara Mertz, *Temples, Tombs and Hieroglyphs* (New York: Delta Publishing Co., 1964).

3. Bjorn Landstrom, *Ships of the Pharaohs* (New York: Doubleday, 1970).

4. Peter Tompkins, *The Secrets of the Great Pyramids* (New York: Harper and Row, 1971).

5. Cheops was constructed almost 5,000 years ago but it was not until 1925, when the base of the pyramid was finally cleared, that the four corners were located with precision by a British engineer, J. H. Cole, and each side was found to be 500 Egyptian cubits, the precise distance traveled by the sun in one half second. While the base of the Pyramid of the Sun at Teotihuacan can no longer be computed to the millimeter, all calculations show, according to Hugh Harleston, that the base is identical to that of Cheops, namely, 500 Egyptian cubits. Speculation on how Cheops was erected continues in the scientific journals of the present day.

6. Tompkins, *Great Pyramids*, pp. xiv and xv.

7. Ibid., p. 17.

8. Barry Fell, *America B.C.* (New York: New York Times Books, 1976), p. 257.

9. Ibid., pp. 253-60.

10. Of course two people separated by time and distance might independently devise the same symbol, or glyph, for gold, considering the complexity of this Egyptian hieroglyph some computer specialist might calculate the probabilities of two individuals devising the identical symbol.

11. Fell, *America B.C.*, pp. 261-67.

12. A number of societies in which Egyptian influence appears have names for a monotheistic god that bears a philological relationship to the word "Aton" used by Iknaton. Sigmund Freud in *Moses and Monotheism* called attention to the similarity of the term "Aton" and the word for God spoken only in the Jewish temple, "Adonai." Note the Micmac "Atnaguna," the Inca use of "Inti," and the Easter Islander's "Atua."

Chapter 9: The Mayan Synthesis

1. Earl H. Swanson, *The New World* (Oxford: Elsevier Phaidon, 1975), pp. 92 et seq.

2. Peter Tompkins, *The Secret of the Great Pyramid* (New York: Harper and Row, 1973), p. 185. "The pyramid of Xochicalco contains a tubular well down which the sun shines perpendicularly without a shadow on a specific day of the year"—evidence that the Mexican pyramid, like Cheops, was an astronomical instrument.

3. Peter Tompkins, *Mysteries of the Mexican Pyramids* (New York: Harper and Row, 1976), p. 172.

4. William Weber Johnson, *Mexico* (New York: Time Incorporated, 1966), p. 49. See also *The Smithsonian*, Washington D.C., May 1978.

5. Swanson, *The New World*, pp. 126-27.

6. Ibid., pp. xvi and xvii.
7. Ibid., p. 98.
8. Ibid., pp. 95-97.
9. Tompkins, *Mexican Pyramids*, pp. 77-80.
10. Ibid., p. 82.
11. Ibid., pp. 110-12.
12. Ibid., p. 112.
13. Samuel A. D. Mercer, *An Egyptian Grammar* (London: Luzac, 1927).
14. Tompkins, *Mexican Pyramids*, p. 111.
15. Ibid., p. 172.
16. Ibid., pp. 292-93.
17. Kingsboro devoted his life to the study of these documents and also the relevant texts in Sanscrit, Hebrew, Greek and Latin. He was not able to escape the evidence of strong Semitic influence in all that he saw and read and suggested that this influence came not from Phoenicia, but from the Lost Tribes of Israel. Ancient Israel and Phoenicia were neighbors and doubtless influenced each other greatly. The capital of Phoenicia was Tyre in present-day Lebanon, a city referred to repeatedly in the Old Testament, and the home port for the vast Phoenician merchant marine. On contract, the Phoenicians built the greatest building of Bible times, namely, Solomon's Temple. They supplied the Hebrews' need for shipping from the time of Moses. See Gerhardt Herm, *The Phoenicians* (New York: Morrow, 1975), pp. 83-124.
18. Modern astronomers have gone over Mayan calculations and discovered that the ancient scholars predicted equinoxes and eclipses up to the year A.D. 26536 to an accuracy of one second. European and American scientists have not been able to produce comparable predictions before the last half of the twentieth century.
19. Tompkins, *Mexican Pyramids*, p. 313.
20. Ibid., p. 253.
20. C.W. Ceram, *Gods, Graves & Scholars* (New York: Alfred A. Knopf, 1952), pp. 318-19. "Not until the 19th Century did the concept of a million become common in the West. By contrast, a cuneiform text was found on the mound of Kunyimjik (Mesopotamia) with a mathematical series the end product of which in our number system would be expressed as 195,955,200,000,000," a figure that would not enter European calculations until the time of Leibnitz or Descartes."
21. Tompkins, *Mexican Pyramids*, p. 253. A report and analysis of Hugh Harleston Jr.'s findings are made by Tompkins pp. 241-81. See also "A Mathematical Analysis of Teotehuacan" by Hugh Harleston, Jr., Mexico City, in *International Congress of Americanists* 41 (October 3, 1974).

Chapter 10: *From Mastery to Oblivion*

1. Hector M. Calderon, *La Ciencia Mathematica de los Mayas*, (Mexico: Editorial Orion, 1966). See also Peter Tompkins, *Mysteries of the Mexican Pyramids*, (New York: Harper and Row, 1966), pp. 286-88. According to Calderon, the Mayan merchant or housewife did not use the dot-and-dash system portrayed in the stone carvings, but rather used two colors of grain, one color for the numeral one, and one for the numeral five. The checkerboard pattern used for calculation became a favorite decorative device and was used in painting, on clothes, and on mats.

2. Earl H. Swanson, *The New World* (Oxford: Elsevier-Phaidon, 1975), pp. 10 et seq.

3. Tompkins, *Mexican Pyramids*, pp. 141-52.

4. Constance Irwin, *Fair Gods and Stone Faces* (New York: St. Martin's Press, 1963); see also Tompkins, *Mexican Pyramids*, pp. 348-49.

5. Jacques Lafaye, *Quetzalcoatl and Guadalupe* (Chicago: University of Chicago Press, 1977), p. 22.

Chapter 11: Montezuma's Testimony

1. Peter Tompkins, *Mysteries of the Mexican Pyramids* (New York: Harper and Row, 1967), pp. 16-17.

2. Ibid., pp. 14-22.

3. Ibid., p. 11. See also Hernan Cortes, *The Conquest of Mexico* (Philadelphia: David Hogan, 1801).

4. Tompkins, *Mexican Pyramids*, p. 11. See also W.H. Prescott, *The History of the Conquest of Mexico* (New York: Harper and Row, 1943).

Chapter 12: Vikings - Moody Adventure

1. Magnus Magnusson, *Viking Expansion Westwards* (London: The Bodley Head Ltd., 1973), p. 114. When the Norse first established colonies in Greenland they came across artifacts left by the Eskimos of North America, such as abandoned boats and evidences of clearing and cultivation. Apparently the Eskimos had settled Greenland and abandoned it before the Norse arrived. However, the early Norse settlers were subject to hostile attacks—whether from Eskimos or Indians is unclear. It is from these people that the Norse apparently secured their first seed corn whose line of supply went back to the Indians of the central plains of Mexico.

Corn is one of the great horticultural triumphs of all times, attributed to the dwellers of the Oaxaca Valley. It is the only plant completely dependent on man for its seeding and, therefore, its survival. The fact that the Norse on Greenland planted corn would be certain evidence of personal contact with the mainland. However, the possibility that previous Indian settlers on Greenland grew their own corn must also be allowed.

2. Smauel E. Morison, *The European Discovery of America* (New York: Oxford

University Press, 1971), pp. 39-40. Morison gives Bjarni credit as being the first "documented" and indubitably European discoverer of America" almost 500 years before Columbus. His great contribution, according to Morison, was to inspire Leif in his colonizing effort. This may be true, but it is undoubtedly also true that Leif knew about the New World before Bjarni's "discovery."

3. Magnusson, *Viking Expansion*, p. 85.

4. Morison, *European Discovery of America*, p. 49.

5. Magnusson, *Viking Expansion*, pp. 147-48.

6. Ibid., p. 128.

7. Morison, *European Discovery of America*, p. 54.

8. Ibid., p. 54.

9. Authorities differ on the size of this colony. It may have been as small as 150 or as large as 250. All accounts agree the shortage of women became a problem, and a source of internal friction. An obvious solution would have been intermarriage with local women. Perhaps racial or religious prejudice prevented this mingling from taking place, although it should be noted that in the history of such settlements, some accomodation of religious and social values is generally devised to permit sexual expression if not intermarriage. Thorfinn's colony seems to be an exception to the general rule that all-male colonies tend to intermarry and as a consequence cultural tensions tend to diminish. Whatever they were, the inhibitions that bedeviled Thorfinn's settlers were not insurmountable for some of the later Scandanavian arrivals.

10. Hear what a modern Viking explorer has to say on this subject. See Thor Hyerdahl, *The RA Expeditions* (New York: Doubleday and Co., 1972) pp. 41-44.

11. Morison, *European Discovery of America*, p. 147-48.

Chapter 13: History and Mystery

1. Samuel E. Morison, *The European Discovery of America* (New York: Oxford University Press, 1971).

2. Hjalmar Holand, *Norse Discoveries and Exploration in America, 982-1362* (New York: Dover, 1969), pp. 64-75.

3. Magnus Magnusson, *Viking Expansion Westwards* (London: The Bodley Head, 1973), p. 141.

4. Compare the Newport tower with Mellifont Abbey, County Louth, built 1142. See Brian DeBreffny, *The Churches and Abbeys of Ireland* (New York: W.W. Norton, 1976).

5. Rejection by the scholarly world is not a unique experience for archeologists, historians and others. A list of examples could be provided. Brasseur de Bourbourg found a Mayan Codex of some 70 pages which he offered to the British Museum and which was returned to him as a forgery. He then offered it to the Royal Library of Paris. They returned it, and fearing the contamination of their standards, refused to keep it either for storage or for study. However, additional portions of the Codex continued to be found. Brasseur, with the indestructible determination of a bloodhound,

continued the search for the missing parts. When all of the 112 pages were assembled the Codex was recognized as authentic, and as the most valuable single Mayan document to become available to the world.

6. Holand, *Norse Discoveries*. All further references to Hjalmar Holand in this chapter are based on this work.

7. C.W. Ceram, *The Secret of Hittites* (New York: Knopf, 1955).

8. He was not alone. Max Planck worked out the foundations of modern mathematics without which the achievement of Einstein would not have been possible. Once, after a mathematical convention, he was asked, "What do you do with the real experts who argue so effectively against you?" To this his reply was, "I just wait for them to die."

9. At its height, the population of Greenland may have been as large as 3,000. If the same logic is used as that which attributed discovery to Columbus, Greenland should be regarded as the first documented European colony in America. Founded in 960, the colony was in the Western Hemisphere, and is on an island closer to the mainland than is Haiti, where Columbus landed in 1492.

10. Nothing can ever diminish the significance of Knutson's voyage of search and discovery. However, Knutson could have made this trip without being out of sight of land for more than a day, and he could expect little surprise in following the coastline of North America, which had been known and explored for more than 350 years. If Knutson, who was on the King's business, sailed directly from Greenland to the western shore of Hudson Bay, he took a trip of approximately 1,000 miles. This would be a memorable undertaking, but not a particularly courageous or frightening one. In the summer of 1978, two college girls rowed an open boat from Seattle to Alaska, a distance of 1,151 miles. The reason for the trip was simply that they "liked to row."

11. Magnusson, *Viking Expansion*, pp. 47, 51, 82, 95.

Chapter 14: Problems of Early America

1. Nigel Davies, *Voyagers to the New World* (New York: William Morrow & Co., 1979) p. 254.
2. Ibid., p. 241.
3. *The Saginaw News*, Jan. 2, 1977.
4. *The Moscow News*, May 25, 1979.

Bibliography

Antiya, Aziz S. *History of Eastern Christianity*. London: Methuen, 1968.

Bainton, Roland H. *The History of Christianity*. New York: American Heritage, 1964.

Calderon, Hector M. *La Ciencia Matematica de los Mayas*. Mexico: Editorial Orion, 1966.

Ceram, C. W. *The Secret of the Hittites*. New York: Knopf, 1955.

———. *Gods, Graves & Scholars*. New York: Knopf, 1952.

Collier, John. *Indians of the Americas*. New York: Norton, 1952.

Davidson, Basil. *The Lost Cities of Africa*. Boston: Little Brown & Co., 1970.

DeBreffny, Brian. *The Churches & Abbeys of Ireland*. New York: W. W. Norton & Co., 1976.

Evans, Estyn. *Prehistoric & Early Christian Ireland*. New York: Barnes & Noble, Inc., 1966.

Fell, Barry. *America B.C.* New York: Quadrangle/The New York Times Book Co., 1976.

Harelson, Hugh, Jr. *International Congress of Americanists*. Vol. 41. October 3, 1974.

Herm, Gerhard. *The Celts*. New York: St. Martin's Press, 1977.

Heyerdahl, Thor. *Aku-Aku*. Chicago: Rand McNally, 1958.

———. *The RA Expeditions*. New York: Doubleday & Co., 1972.

Holand, Hjalmar. *Norse Discoveries and Exploration in America, 982-1362*. New York: Dover Publications, Inc., 1969.

Homer. *The Odyssey*. Translated by Albert Cook. New York: W. W. Norton, 1967.

Howard, Cecil. *Pizzaro and the Conquest of Peru*. New York: Harper & Row, 1968.

Irwin, Constance. *Fair Gods and Stone Faces.* New York: St. Martin's Press, 1963.

Johnson, William Weber. *Mexico.* New York: Time Incorporated, 1966.

Lafaye, Jacque. *Quetzalcoatl and Guadalupe.* Chicago: University of Chicago Press, 1977.

Landstrom, Bjorn. *Ships of the Pharaohs.* Garden City, N.Y.: Doubleday, 1970.

Magnusson, Magnus. *Viking Expansion Westwards.* London: Bodley Head Ltd., 1973.

Mercer, Samuel. *An Egyptian Grammer.* London: Luzac, 1927.

Mertz, Barbara. *Temples, Tombs and Hieroglphys.* New York: Dell Publishing Co., 1964.

More, Sir Thomas. *Utopia.* New York: Washington Square Press, 1965.

Morgan, Arthur. *Nowhere Was Somewhere.* Chapel Hill, N.C.: University of North Carolina Press, 1946.

Morison, Samuel Eliot. *The European Discovery of America, The Northern Voyages, U.S.A., 500-1600 A.D.* New York: Oxford University Press, 1971.

Mumford, Lewis. *The Condition of Man.* New York: Harcourt Brace & Co., 1944.

Prescott, William H. *Conquest of Peru.* New York: Hurst and Company, 1847.

The History of the Conquest of Mexico. New York: Harper & Row, 1843.

Ross, Anne. *Pagan Celtic Britain.* London: Sphere Books Ltd., 1974.

Ryan, John, S.J. *Irish Monastascism.* Ithaca, N.Y.: Cornell University Press, 1972.

Sagon, Carl. *The Dragons of Eden.* New York: Random House, 1977.

Severy, Merle, *"The Celts." National Geographic,* October, 1977.

Swanson, Earl H. *The New World.* Oxford: Elsevier Phaidon, 1975.

Toeffler, Arnold. *Future Shock.* New York: Random House, 1970.

Tompkins, Peter. *Mysteries of the Mexican Pyramids.* New York: Harper & Row, 1976.

———. *The Secret of the Great Pyramid.* New York: Harper & Row, 1973.

Warren, F. E. *The Liturgy & Ritual of the Celtic Church.* Oxford: Clarendon Press, 1881.

Weiner, Leo. *Africa and the Discovery of America.* Philadelphia: Innes and Sons, 1920.

Index

Abyssinians, 134
Acadia, 125
Aeschylus, 64
Africa, 12, 15, 17, 23-26, 48, 50, 74
    101, 112-115, 130, 135, 182
Ailbe, 57
Alaric, 28, 36, 46-49
Alaska, 15
Aleutians, 15
Alexandria, 41, 62
Algonquins, 124, 127, 128
Althing, 177
Amish, 104
Andes, 74, 80, 86, 92
Annabaptist Brethren, 104
Antilles, 23
Arctic Circle, 176, 181
Arian, Arius, 41-45, 50, 55
Armagh, 55
Arnarson, Ingolf, 176
Arnold, Benedict, 188
Arthur, King, 39
Aswan, 120
Atahualpa, 86, 92, 94-97
Atlantis, 122
Atlas Mountains, 50
Atnaquna, 128
Aton, 35, 128
Augustine, 41, 55
Australia, 136
Azores, 113
Aztec, 62, 73, 89, 90, 137, 151

Babylonians, 148
Bainton, Roland H., 50, 52
Balthi, 46
Baltic Sea, 114
Bantu, 15

Barinth, Abbott, 56, 57
Beltone, Day of, 39
Bering Straits, 67, 137, 211
Bianco, Andrea, 75
Bingham, Hiram, 77
Bithnia, King of, 40
Black Hills, S. Dak., 132
Blattahlid, 180
Book of Ballymote, 34
Book of Kells, 37
Boole, 159
Borglum, Gutzon, 132
Bosporus, 40
Bothnia, Gulf, 114
Boudicca, 38
Bourbourg, Abbe, 139, 140
Brandeis University, 116
Brazil, 217
Breda, O. J., 191
Brendan St., 56, 57, 58, 104, 115
Brittany, 28, 56
Bruderhoff, 104
Burgundy, 50
Byzantium, 40, 176

Cabera, Dr. Paul Felix, 139
Cabinet d'Egyptologic, 64, 145
Cadavaco, 216
Cadiz, 29, 31, 35, 100, 112
Caesar, Julius, 52, 65
Cairo Museum, 70
Cajamarca, 94
Calvinism, 51
Cambodia, 213
Canary Islands, 112, 115
Carausius, 176
Caribbean, 75, 101

241

Carlotta, Empress, 31
Carthage, 49, 110, 114, 115, 139, 144
Caspian Sea, 176
Catlin, Geo., 206
Celts, 28, 29, 30, 33, 34, 36-41, 45-49, 50-55, 87, 101, 127, 132, 144, 176, 177, 182, 189, 190
Ceram, C. W., 148, 194
Cerne Giant, 31
Chalcedon, Council of, 43, 144
Challcochima, Gen., 95
Champollion, 125, 127, 141
Charles V, 93, 96, 169
Charney, Claude, 162
Cheops, 121, 122, 149, 153
Cherokees, 124
Chiapas, 138
China, 12, 13, 16, 17, 215
Chitzen Itza, 134
Choc-Mool, 147
Cholula, 170
Chomsky, Noam, 71
Christmas Tree, 39
Clare County, 34
Clark, Fort, 206
Clark, Lord Kenneth, 45
Cleopatra, 125
Clovis, 50
Coaque, 93
Cod, Cape, 175, 179
Collier, John, 72
Cologne, 12
Columba, 53, 55
Columbus, 15, 18, 23-26, 75, 110, 119, 168, 202, 218-219
Columbcille, 55
Constantine the Great, 41, 42
Constantinople, 43, 47, 48
Constantius, 42
Copernicus, 18, 19, 112, 119
Coptic Church, 13, 144
Coptos, 118, 123
Coraquenque, 86
Corinth, 123
Cornwall, 114
Cortes, 62, 73, 124, 129, 141, 153, 155, 164, 166, 168, 194
Corvo, Island of, 113
Crete, 62, 63
Cuba, 24
Curme, Prof. Geo., 192
Cuzco, 85, 86, 100

Davenport, Iowa, 126, 127
Davies, Nigel, 220-221
Denmark, 176, 177
Dido, Queen, 110
Dighton, 34
Dingle, 34
Dodd, James E., 184, 185, 186
Donatists, 49
Douglas County, 190
Douglas, E. D., 65
Druid, 39, 44, 53

Easter Island, 67, 147, 215
Ecuador, 70
Egypt, 13, 14, 17, 53, 63, 64, 74, 87, 100, 117-28, 132-35, 144, 155
Einstein, 18, 153
Epstein, Jerry, 222-23
Eratosthenes, 120
Erickson, Leif, 178, 180, 181, 187
Eric the Red, 60, 173
Erie, Lake, 200
Erikson, Erik, 81
Eriksson, King Magnus, 199, 203
Eskimos, 211
Ethiopia, 13, 24, 118
Ethnobotanical Laboratory, 69
Euclid, 161

Faroe Islands, 56, 182
Fell, Barry, 29, 30, 33, 36
Ferdinand, 24
Fermi, Eurico, 18
Flatey, Jarbok, 174, 179
Florida, 23

France, 43
Freydis, 38

Gadansk, 114
Galatia, 40
German, Germany, 18, 25, 41, 44
Gibralter, 115, 136
Godfrey, Wm. S., 188
Gordon, Cyrus H., 116
Great Lakes, 200
Greece, Greeks, 14, 47, 65
Greenland, 26, 56, 58, 173-77, 180, 181, 182, 199
Guanin, 23
Guatemala, 139
Gudrid, 180, 182
Guinea, 23, 24, 25, 26

Hadrians Wall, 35
Haiti, 23
Halloween, 39
Hanno, Captain, 115, 116, 124, 126
Hapshepsut, Pharaoh, 118, 124
Harelson, Hugh Jr., 150, 152, 153
Harrington, James, 104
Harvard University, 75, 188, 202
Heine-Gelden, Robert, 216
Herjolfsson, Bjorni, 173, 174, 187
Herodotus, 112
Hesiod, 114, 115
Heyerdahl, Thor, 67, 68, 71, 74, 86, 113, 117, 118, 135, 198
Hittites, 13
Huaca del Sol, Pyramid, 133
Hudson Bay, 200
Humboldt Current, 136
Holand, Hjalmar, 192-202, 221-2
Homer, 109, 110, 115
Hythloday, 74, 75

Iceland, 56, 173, 174, 176, 180-82
Ikhnaton, 11, 35, 36, 63, 128
Iliad, 56, 60

Index • 243

Inca, 17, 67, 71-;79, 82, 84, 86, 90-95, 105
Ingstadt, Helge, 178, 179, 183
Inti, 84
Ireland, 29, 31, 35, 37, 39, 41, 43, 45, 50, 54, 58, 101, 176
Irwin, Constance, 148, 164
Israel, 14, 46, 134

Jacob, John, 185
Japanese, 215
Jerome, Saint, 42
Jesus Christ, 43, 97, 104, 167, 168, 170
Jews, 46
John II, King, 27
Johnson, John, 128
Juan, King, 23, 25
Justinian, 49, 55

Karlsefni, Thorfinn, 180, 182
Kensington Stone, 191, 95
Kerry, County, 31, 34
Kiev, 46, 176
Kingsboro, Lord, 143
Knorosov, 143
Knossos, 62, 63
Knudtzon, 63
Knutson, Paul, 200-203

Labrador, 178
Lands End, 114
Landstrom, Bjorn, 118
L'Anse aux Meadows, 178, 183
Lear, Edward, 143
Levi-Strauss, Claude, 216
Lhyd, Edward, 34
Libby, Willard, 66
London, Roman, 38, 56
Lorillard, Pierre C., 162
Lutheranism, 51
Luxor, Temple of, 123

Before Columbus

Madagascar, 12
Magellan, Strait of, 67
Magnusson, Magnus, 68, 176, 189, 209
Maillard, Pierre, 124, 125, 126, 127
Maine, 127
Mamum, Abdulla Al, 122
Man, Isle of, 179
Manchu Picchu, 77
Mandan Indians, 204-10
Maori, 216-17
Markland, 178
Marque, M. de la, 205
Marsailles, 100, 112
Mary, 43
Massachusetts, 104
Mather, Cotton, 34, 35
Maximilian, Emperor, 31
Maximilian, Prince, 206, 209
Maya, 65, 120, 124, 129-53, 154-64
Maypole, 39
Mediterranian, 110
Mendoza, 86
Menes, Pharaoh, 11
Mennonites, 104
Mercer, Samuel A. D., 141
Mernoc, 56, 57
Mexico, 15, 31, 33, 62, 68, 116, 120, 129
Michigan, University of, 69, 70
Micmac Indians, 124, 125, 126, 128, 137
Million, Rene, 152
Minnesota, 190
Minnesota, University of, 191
Minot, 204
Mixtec, 137
Moche, 133
Monco Capac, 86
Mongols, 43, 212
Monophysite, 43
Montezuma, 139, 168-72
Moors, 161
More, Sir Thomas, 73, 74, 75, 76, 77, 79, 89, 103
Moreno, Dr. Jimenez, 164
Morgan, Arthur, 77, 84, 86

Morison, Samuel Eliot, 19, 26, 57, 113, 116
Mormons, 104
Morocco, 50, 109, 134
Moses, 11
Moslems, 13
Moundsville, W.Va., 35
Mumford, Lewis, 51
Mycenae, 60
Mystery Hill, 29

Nahuatl, 139, 144
Napoleon, 125, 135
National Observatory, 147
Necho, Pharaoh, 112, 123
Negrito, 213
Negro, 23, 24, 130
Nelson River, 200, 201
Nestorians, 13, 43, 44
New Economy, 104
Newfoundland, 38, 175, 178, 179
New Jerusalem, 104
Newport Tower, 187
Newton, Isaac, 161
New Zealand, 214
Niagara Falls, 200
Nile, 14, 112, 117, 118, 130, 135
Nipigon, Lake, 184, 187
Normandy, 177
North Dakota, 204, 205
Norway, 38, 176-78, 200
Nova Scotia, 175
Novosibirsk Science Center, 143
Nycene Creed, 43

Oaxaca, Valley of, 129
Oceana, 104
Oddson, Bishop, 198
Odyssey, 56, 60
Ogam, 30, 31, 33, 34, 35, 36, 55
Ohman, Olaf, 190, 192
Oklahoma, 35
Olmec, 70, 129, 130, 136, 137, 162, 217

Ontario, Lake, 200
Ordoney y Aquiar, 138
Origen of Alexandria, 52
Owsy, King, 55

Palenque, Temple, 153
Paleo-Botany Research Center, 202
Panama, 92, 93
Paraguay, 35, 74, 217
Patrick, St., 54, 56
Paucartambo, 86
Paul, Apostle, 40, 45, 109, 123
Penn, Wm., 18
Pennsylvania, 104
Pergamon, 40
Persepolis, 162
Persians, 134
Peru, 67, 73-75, 84, 88, 104-07
Peten, Lake, 140
Philadelphia, 18
Phoenicians, 25, 31, 35, 100, 109-10 117, 126, 134, 138, 155, 164
Piaget, Jean, 81
Pizarro, 73, 86, 91-95, 97, 98
Plongeon, Le, 130, 143
Pluto, 145
Polynesian, 215-]17
Port Arthur, 184, 185, 186
Portugal, 101
Portuguese, 25, 26
Prescott, Wm. H., 75-79, 84, 107
Prester, John, 24
Puna, Island of, 93
Punic, 127
Punic wars, 110
Punt, 118
Putnam Museum, 127

Quakers, 104
Quetzalcoatl, 165, 166, 218
Quiche, 139, 141
Quipu, 100

Race, Cape, 113
Ramos, Bernardo, 116

Rasles, Father S., 126
Regensburg, 46
Reinert, Alec, 35
Reis-Batallia, 26
Rhine, 109
Rhode Island, 182, 187
Richelieu, Cardinal, 125
Robinson, E.A., 39
Rochester, University, 152
Rome, Roman, 14, 18, 36, 40, 45, 46, 51-54, 87, 110, 176, 181
Rosetta Stone, 125
Ross, Anne, 34
Round Rock, Texas, 222
Royal Ontario Museum, 185
Russia, 40, 199

St. Lawrence Seaway, 200
Saginaw, Treaty of, 72
San Salvador, 15
Santiago, 86
Saquara, Pyramid of, 121
Savoy, Gene, 77
Scandinavia, 26, 41
Schliemann, H., 60, 162
Schoolcraft, H.R., 124
Schurtz, Karl, 18
Scilly, Isle of, 114
Scotland, 28, 55, 56
Sechura Desert, 92
Seine, 176
Shakers, 104
Shakespeare, 40
Shih Chi, 216
Sierra, 80, 81
Skraelings, 38, 39
Smith, Adam, 78
Snorri, 181
Solmon Indians, 216
Sothic year, 145
Spain, 29-31, 35, 43, 50, 55, 101, 114
Stilicho, 47
Stonehenge, 30, 189, 190
Strabo, 40
Sun, Pyramid of the, 152

Sweden, 200
Switzerland, 88

Tabasco, 138
Talca, 86
Tarshish, Ships of, 109
Tel El-Armarna, 63
Tennyson, Alfred Lord, 115
Tenochtitlan, 62
Teotihuacan, 65, 66, 68, 70, 129, 133, 137, 149, 150; 151, 153, 164, 170
Tertullian of Carthage, 52
Tezcatlipoca, 165
Thebes, 63
Theodosius I, 42, 49
Theodosius II, 43, 48, 49, 55
Thessalonica, 43
Thjodhilds Church, 187
Thomas, St., 167
Tierra del Fuego, 15, 137
Titicaca, 16, 74, 86
Toeffler, Arnold, 108
Toltec, 129, 152, 164
Tompkins, Peter, 119, 120, 130, 148, 149, 153
Totora reed, 16, 17, 71, 75, 118
Touro Park, 188
Trinity College, 37
Tristram, 39
Troy, 60, 109
Tula, 162, 164, 165
Tumbes, 92, 93
Tutankhamen, 12
Tutishcainyo, 70
Tykir, 179

Ucayali, River, 70
Ulysses, 104, 115
"Uncle Sam," 164
United States, 33, 79, 114
U.S. Supreme Court, 127, 134
Utopia, 73, 75
Uxmal, 31

Vallancey, C., 34
Valverde, Friar Vincete de, 96
Venezuela, 25
Venta, La, 130, 164
Verde, Cape, 15, 75
Verendrye, Capt. Pierre la, 204, 205, 206
Vespucci, Amerigo, 75
Vikings, 51, 60, 179, 184
Vilcabamba, 77
Vineland, 180
Viracocha, 86, 90, 91, 92, 96, 98, 99, 102, 137
Visigoths, 50
Volga, 176
Votan, 138
Vrusa, Gen. Martin de, 140

Wadi Hammadi, 123
Wagner, Richard, 39
Wales, 28, 56
Washington Monument, 133
Weiner, 26
Widener Library, 30, 125
Winnipeg, Lake, 201
Wright, Quincy, 54

Xiknalkan, 145

Yucatan Peninsula, 66, 129, 133, 134, 135, 136, 138

Zaire, 13
Zambezi, 12, 15
Zoyzer Pyramid, 130